ON TRIUMPH

1962-1967

Reprinted From
Cycle World Magazine

ISBN 1 869826 574

Published By
Brooklands Books with permission of Cycle World

Titles in this series

CYCLE WORLD ON BMW 1964-1973
CYCLE WORLD ON BMW 1974-1980
CYCLE WORLD ON BMW 1981-1986
CYCLE WORLD ON HARLEY-DAVIDSON 1962-1968
CYCLE WORLD ON HARLEY-DAVIDSON 1968-1978
CYCLE WORLD ON HARLEY-DAVIDSON 1978-1983
CYCLE WORLD ON HARLEY-DAVIDSON 1983-1987
CYCLE WORLD ON TRIUMPH 1962-1967
CYCLE WORLD ON TRIUMPH 1967-1972
CYCLE WORLD ON TRIUMPH 1972-1987

Titles in preparation will cover: BMW, BSA, Ducati, Kawasaki, Honda, Husqvarna, Norton, Suzuki, Triumph, Yamaha, etc.

Distributed By

Cycle World
1499 Monrovia Avenue
Newport Beach
California 92663 U.S.A.

Brooklands Book Distribution Ltd.
Holmerise, Seven Hills Road,
Cobham, Surrey KT11 1ES,
England

CYCLE WORLD

5	1962 Triumph Bonneville Road Test	*Cycle World*	Jan.	1962
10	Triumphs for 1962	*Cycle World*	Jan.	1962
12	Triumph T-100 SR Road Test	*Cycle World*	August	1962
16	Triumph for 1963	*Cycle World*	May	1963
17	Triumph T-120 TT Special Road Test	*Cycle World*	May	1963
22	Triumphs Trials Trail Test	*Cycle World*	August	1963
23	We Triumph at Bonneville	*Cycle World*	Nov.	1963
26	Something for a Starter	*Cycle World*	March	1964
28	H&C Triumph Special	*Cycle World*	June	1964
31	Triumph Mountain Cub Trail Test	*Cycle World*	July	1964
32	Triumph Thunderbird Road Test	*Cycle World*	August	1964
36	We Try at Bonneville	*Cycle World*	Nov.	1964
38	"Putting its Best Wheels Forward"	*Cycle World*	Oct.	1964
43	Three-Way Triumph	*Cycle World*	May	1965
44	Triumph TR-6 SR Road Test	*Cycle World*	May	1965
48	Dick Rios' Two-Timer	*Cycle World*	Sept.	1965
50	Triumph TR6SC Trophy Special Road Test	*Cycle World*	Oct.	1965
54	Triumph T120/R Bonneville Road Test	*Cycle World*	Jan.	1966
58	Triumph History	*Cycle World*	March	1966
67	A Most Versatile Motorcycle	*Cycle World*	May	1966
70	Hagon-Triumph Dragster Road Test	*Cycle World*	Oct.	1966
73	Triton-Vee Four	*Cycle World*	June	1966
74	Triumph 500 Road Racer Road Test	*Cycle World*	Feb.	1967
78	A Pair of Desert Sleds	*Cycle World*	Nov.	1967

We are frequently asked for copies of out of print Road Tests and other articles that have appeared in Cycle World. To satisfy this need we are producing a series of books that will include, as nearly as possible, all the important information on one make or subject for a given period.

It is our hope that these collections of articles will give an overview that will be of value to historians, restorers and potential buyers, as well as to present owners of these interesting motorcycles.

COPYRIGHT © 1987 by CBS MAGAZINES, A DIVISION OF CBS INC.

Printed in Hong Kong

1962 TRIUMPH BONNEVILLE

Cycle World Road Test

DESPITE the present-day trend toward small-displacement motorcycles, there is now and will very likely always be a sizable group of riders for whom nothing short of a genuine over-100-mph machine is ever enough. These are the people who really enjoy the heady thrust that only comes with relatively large, powerful engines. On the other hand, most of these men are experienced enough to recognize and appreciate the merits of light weight and compact overall dimensions, and they like to have these features as well. Not many years ago, to have both the power **and** agility was quite impossible, but progress — and the Triumph Motorcycle Company — has brought us a machine that possesses a surprisingly large measure of each: the Bonneville T120.

In its essentials, the new T120 is a very close development of preceding Triumph twins. The vertical-twin cylinder layout — which has proven, over the years, to be worth imitating — is naturally retained. The crankshaft still runs in two bearings (the one on the output side is of the roller type), and retains the large, central flywheel. The lack of a center mainbearing may be bothersome to a few, but constant detail improvement in the crank, its bearings and supporting structure has kept the engine's overall reliability well apace with the frequent increases in output.

At the top of the engine is the deep-breathing, splayed-port cylinder head, cast in light-alloy, like all of those in the rest of the Triumph line, and presenting quite an impressive companion piece to the two Amal monobloc carburetors. On our test machine, these used the conventional, integral float chambers — which now have an inter-connecting fuel-feed line, to permit the use of standard fuel taps in the reserve system. Available, too, for those who cannot be satisfied with the Bonneville's already staggering performance,

is an optional remote-float setup. This relieves the normal float chambers of their duties entirely, and overcomes a slight fuzziness at about 6800 rpm. This is caused by vibration-induced frothing in the stock float chambers and upsets the mixture a trifling amount at that particular engine speed. We were not overly concerned with the frothing problem, as the Triumph's performance was, to be absolutely honest, about as much as we could comfortably handle anyway.

As was the case on the engine, there was no faulting the workmanship on the rest of the machine. There are new colors for this year's — which were very attractive — and everything that could conceivably need to be polished or chromed had received the full treatment. As it goes almost without saying, the brightwork contributes nothing to performance, but it is invaluable in producing and maintaining pride of ownership, and we think that is certainly worth something.

The Bonneville color scheme for 1962 really has "it." A reddish burnt-orange is used on the upper side of the fuel tank; and, on the lower side, less prominently, a medium-grey that contrasts just nicely. Both paints are of the metallic variety and are beautiful examples of the type. The metallic quality is not done so heavily that it pokes you in the eye; and yet, there is enough to give the finish that "inch-deep" look that is understandably so popular. The fenders are of the same grey as that used on the lower part of the tank; and, for trim, there is an inch-wide strip of the orange running down the middle. The final touch is provided by a thin line of gold on each side of the orange, separating it from the grey. In overall appearance, the finish hovers somewhere between studied elegance and eye-catching flash — you can just take your pick.

The T120 frame and suspension system follows the pattern that has become a standard. There is a two-loop tubular frame cradling the engine and transmission (this last item having dual adjusters) and carrying telescoping, oil-damped forks at the front and trailing links astern. The combination spring-damper units at the rear are, as is usually the case, adjustable and can be set to whatever the load carried requires. We might mention that the springing suited us exactly: Stiff enough to help the rider in maintaining control throughout the Bonneville's considerable speed range, and yet well short of harshness.

During the time we had the T120, it started with heartening ease. There was no provision for choking — beyond the usual float "ticklers" — but a mild pre-flooding would get the engine fired up quickly every time. It starts easily despite the engineering pitfall inherent in having 50 bhp from 40 cubic inches. And, although the idle was a bit lumpy (as one might well expect) there was no problem with those embarrassing mid-intersection "conk-outs". The engine further belied its high power output by its uncommon willingness to idle-down in 4th and plug along at low speeds. From about 2000 rpm, in 4th, one can just crank it on; and, as long as too much throttle is not given until the speed builds up, the T120 will pull smoothly away. Not only that, but with even our fairly bulky test rider aboard and sitting bolt-upright, this remarkable machine would easily push past the 100 mph mark and run steadily at that velocity as though it would never tire.

More vigorous techniques had their effect, too, make no mistake about that. Using 7000 rpm as a shifting point, we were able to snatch the T120 up to a bit over 85 mph in the standing 1/4-mile, and with an elapsed time of 14.5 seconds. Inasmuch as our test rider spent a sizeable portion of the "quarter" fighting to keep the front wheel down and the machine pointed, we are of the opinion that a really top-flight rider might best even that excellent time. Certainly, the bike itself was willing enough.

The clutch engages without any of that distressing suddenness sometimes found in powerful motorcycles, and when "all-in" showed no trace of slippage. The only difficulty was one we rather liked: the Triumph has so much power that it creates something of a situation when the rider is trying for all-out performance from a standing start. Given sufficient time to completely familiar-

PERFORMANCE

Top speed, average	108
best run	110
Max. speed in gears (7000 rpm)	
3rd	94
2nd	66
1st	46
Fuel consumption range, mpg	40/70
Mph per 1000 rpm	16.0

SPEEDOMETER ERROR

30 mph, actual	29.0
50	47.7
70	65.6
90	83.8

ACCELERATION

0-30 mph, sec.	1.9
0-40	3.0
0-50	5.1
0-60	6.5
0-70	9.1
0-80	12.2
9-90	17.0
0-100	27.0
Standing 1/4 mile	14.5
speed reached	85

SPECIFICATIONS

List price	$1,139.00 F.O.B., P.O.E.
Frame type	tubular, 2- loop
Suspension, front	telescopic fork
Suspension rear	trailing links
Tire size, front	3.25-19
Tire size, rear	4.00-18
Brake lining area, sq. in.	32.5
Engine type	vert. twin
Bore & stroke	2.79 x 3.23
Displacement, cu. in.	36.5
Displacement, cu. cent	649
Compression ratio	8.5:1
Carburetion	(2) 1 1/16 Amal monobloc
Ignition	Lucas Magneto
Bhp @ rpm	50 @ 6500
Fuel capacity, gals.	3.5
Oil capacity, pts.	6.0
Oil system	dry sump
Starting system	kick, folding crank

POWER TRANSMISSION

Clutch type	multi-disc, oil-bath
Primary drive	single-row chain
Final drive	single-row chain
Gear ratios, overall:1	
4th	4.9
3rd	5.8
2nd	8.3
1st	11.9

DIMENSIONS, IN.

Wheelbase	55.2
Saddle height	31.5
Foot-peg height	11.5
Ground clearance	5.0
Saddle width, max.	11.0
Curb weight, lbs.	410

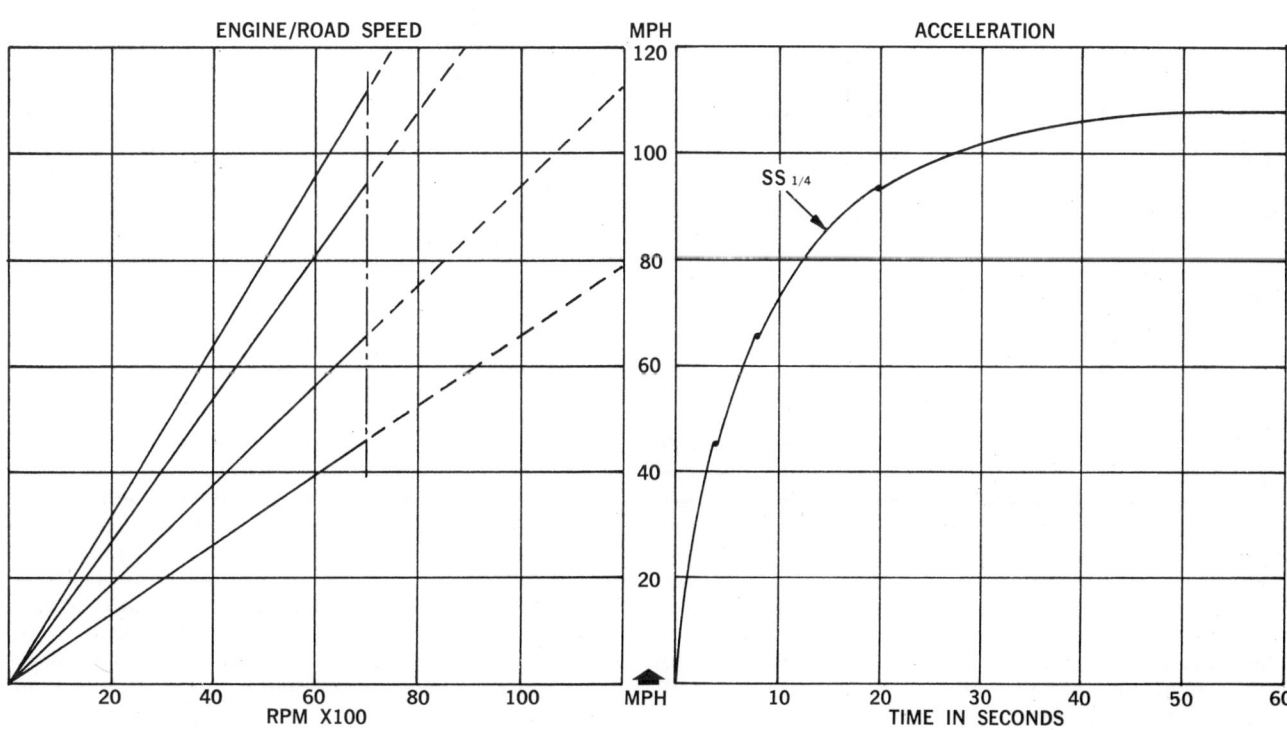

ize ourselves with this vast and unaccustomed surplus of power, we could have done our tests with less fuss — but it probably wouldn't have been as much fun.

Apart from this necessity for adjusting to the great amount of power on tap, the rest of the test mileage was covered without incident. Moreover, the miles were also traveled in great comfort. The Triumph is agile, as we previously intimated, and the saddle — besides looking good—was comfortable. We were delighted, too, to find that the positioning of the pegs, relative to the saddle and the bars, enables one to assume a variety of riding positions without getting that out-of-balance sensation. On an extended trip, a seemingly minor point such as this assumes very important proportions.

Other important "minor" points were well cared for, too. Vibration is always something of a problem where motorcycles are concerned, and on the T120 vibration-damping rubber mounts are used extensively. Most noticeable of the rubber mountings was the fuel tank, which is really cradled in the stuff, and on the instruments. Nothing is more exasperating than having fine, accurate instruments (and the Bonneville's chronometric speedo and tach qualify as such) that vibrate so badly that they cannot be read. In the case of the Bonneville, this is prevented by having the instruments fixed on rubber biscuits.

To be absolutely accurate, though, there wasn't much vibration to worry about. We did a lot of just plain medium-speed road cruising; and, at 60-70 mph, the rider is almost as comfortable and unshaken as if he were sitting at home — and he's having a much better time. The saddle is broad and soft (enough for two) and the springing is, as we said, just the thing. And, there is a lot of mental comfort, while passing cars, in knowing that hooking on a big handful of throttle will ram you ahead like something going into orbit.

Touring through winding mountain roads, and feeling perhaps too much like the legendary John Surtees for our own continued well-being, we tried a bit of what the British call "ear-holing". As one might guess, this involves leaning into the corners at such speed and with such purpose that one's ear-hole does, in fact, get down pretty near the pavement. Thus engaged, a good idea of the T120's handling soon emerged, and it is for the most part very impressive. Only on fast, cambered bends, where the added loads imposed by centrifugal force begin to overwork the suspension did anything unsettling develop. There, particularly when the road turned choppy, the suspension damping was to some extent overcome, and the bike would feel a trifle uncertain. However, this may be said of most road bikes, for one cannot have a touring ride and Grand Prix-caliber cornering all in the same machine.

Stopping is important, too, and the Triumph's capabilities in that area were equal to its performance. The brakes are not overly large, compared to some we have seen recently, but there is no doubt that they do the job. The floating-shoe brakes introduced last year are continued on the new model and the actuating system (the cable and housing, specifically) has been improved to eliminate the faint springiness present previously. The same treatment has been accorded the clutch, with the result that those handle-bar levers now feel more positively linked to their respective components than before.

There are other minor changes, such as the revised fuel taps now being used; and the electrical system has been revised to meet the specific requirements of each model in the Triumph line—the T120 being no exception. All in all, though, the Bonneville for 1962 is the same machine as before: extremely fast and yet docile, and as desirable a performance package as it has ever been our pleasure to ride.

, etc.....

RACING LUCK — One can only speculate as to the thoughts churning through Joe Leonard's mind at this moment, but to be sidelined by a flat tire when leading the Sacramento 20-mile National Championship would seem the bitterest kind of racing disappointment.

PHOTO BY WALT MAHONY

THUNDERBIRD 6T

SPORTS TIGER CUB T20

BONNEVILLE T120

T100 SPORTS

SPEED TWIN 5T/A

TIGER CUB T20

TROPHY TR6

TRIUMPHS FOR 1962

The 1962 range of Triumph has been consolidated to embrace ten models especially suited to U.S.A. requirements. New polychromatic finishes and technical refinements highlight the new line.

Last year, Triumph introduced sixteen models: Five 200cc lightweight Tiger Cub models; Five 350cc and 500cc mediumweight models; and six 650cc medium heavy models. With all of the sixteen 1961 models being consolidated into ten models for 1962, it should help to eliminate confusion arising from having too many models similarly equipped.

The 1962 range is divided into three categories; The A range consisting of two lightweights, the B range embracing five medium heavyweight models, and the C range encompassing three mediumweight models.

"A" RANGE, TWO MODELS TIGER CUB T20

Engine 200cc o.h.v., 10 b.h.p. at 6,000 rpm, engine and 4 speed, foot operated gear box integrally. New type Amal 32/2 carburetor is standard. A new and improved cast iron oil pump has been added. 90 to 100 mpg. Hydraulically controlled front and rear suspension, equipped with Dunlop street tires. Lucas battery-coil ignition. 5 hp model available for use in States with regulations governing 14 year old operators.

SPORTS TIGER CUB T20's

Engine same as Tiger Cub with addition of special camshaft, higher compression and larger carburetor, rated at 15.5 hp. Features 19" front and 18" rear wheels for added ground clearance for use in rough territory. Front suspension is heavy duty and is the same as on the C range models. Electrical system operates under energy transfer (no battery) and direct lighting. Optional equipment includes sports or trials tires, tachometer, special lighting and exhaust systems. Standard equipment includes a dry element air filter; technical improvements include the new oil pump and carburetor as on the T20.

"B" RANGE, FIVE MODELS
THUNDERBIRD 6T

650c ohv vertical twin, alloy cylinder head. Virtually the same as 1961 with new colors available. New features on the Bonneville, as well as the entire "B" range are: A new tape angle fuel outlet sealed with "0" rings, an electrical system employing a closely regulated alternator in separate variations for magneto-alternator and AC coil ignition, stiffer heavy duty clutch and brake cables. The Thunderbird utilizes a new special two-in-one exhaust system on the right side with a new and larger muffler.

BONNEVILLE T120

The Bonneville models are now offered in two dual carburetor versions, one for highway use and the other for competition. The T-120R, the highway version, is capable of 110 mph and features a tachometer as standard equipment. The new T-120C varies only slightly from the highway model in that it is equipped with Dunlop Trials Universal tires and an upswept dual exhaust system for racing.

TROPHY TR6

The third machine in the Triumph "B" range is identical to the Bonneville model in appearance except that it mounts a large bore single Amal Carburetor for more flexibility.

"C" RANGE, THREE MODELS
SPEED TWIN, 5T/A

Ignition in the famous low silhouette Speed Twin is by the new Lucas AC type with battery and coil together with fully rectified lighting. A 350cc version is available on special order. New technical features on the Speed Twin, as well as the other bikes in the "C" range are: A new Lucas ammeter and the continuation of the improved energy transfer (batteryless) igition; a new exhaust camshaft arrangement for the T-100 for more efficient high speed operation; new and stiffer control cables.

T100 SPORTS

Two models of the Sports Tiger 100 are offered for 1962. The T100S/R and T100S/C are both powered by the new unit construction over-square (short stroke) engine developing almost 40hp at 7,000 rpm. It is identical to the Speed Twin engine except that sports camshafts, high compression and a larger carburetor boost its performance capabilities.

The T100S/R is the road model with battery AC coil ignition and rectified lighting, it also utilizes a larger fuel tank and a downswept exhaust system. The T100S/C varies from the road model in that it is equipped with AC energy transfer ignition and direct lighting, Dunlop Trials tires, a smaller capacity gas tank and a competition style two-into-one exhaust system. Both the T100S/R and the T100S/C employs 19" front and 18" rear wheels with sports front and rear fenders.

Cycle World Road Test
TRIUMPH T-100 SR

FOR MANY A YEAR, one of the most popular models in the fast-selling Triumph line has been the "Tiger 100," their 500 cubic centimeter vertical twin. Sports riders, engaging in many and sundry forms of competition riding, liked it because it met the 500cc limit often imposed on engine size and was, more importantly, capable of winning almost any race in which it was entered. Street riders, many of whom had not the slightest notion of going racing, liked it too, because it is very smooth and although lacking the low speed torque of the big, 650cc series Triumphs, the Tiger 100 has always been a spirited performer.

Now, for 1962, there is a new model, called the "Sports Tiger" T100S. The name is very well chosen, because the new machine is exceedingly sporting in concept and a real tiger by nature. This was, of course, also true of the earlier Tiger and the new one is simply more of the same only better.

The Sports Tiger's engine is the end product of many years of polishing and refining on the original long-stroke "Speed Twin." However, the bore is now much larger and the stroke shorter than was once the case. Actually, it is a case of one end of an engine being upgraded to meet the standard set by the other end. Since the beginning, the Triumph twin's valve gear has been willing to work at speeds that had the old long-stroke engine in a complete frazzle. With the introduction of the short-stroke engine, piston speeds, even at the crank speed at which valve bounce occurs, are

moderate. What this means, in practical terms, is that the new series Triumph 500cc engines can be twisted harder than ever before without too much risk of a big bang. In fact, this new engine is smooth, all the way up to the limit imposed by the strength of the valve springs. Until the limit is reached, the engine never feels rough, and the power does not seem to drop off too badly. Our test bike did not come equipped with a tachometer, although that instrument is available on the bike as an extra-cost option. We would recommend very strongly that any purchaser of the Sports Tiger make the extra outlay of cash required to get the tachometer. While we did not break the test bike's engine through our repeated excursions into the area near valve float, it does seem reasonable that overall reliability would be improved if one could always shift before the onset of valve clatter. The Triumph is just too willing for its own good and we think the tachometer is essential.

It is interesting to note that by the use of Webco or Jomo racing valve springs the rpm limit is being increased considerably by many of the prominent Triumph competitors, of which, incidentally, there are a great number.

As in all new-series Triumph 500cc twins, the engine and transmission are in a single unit. This is becoming standard practice for many companies, but unlike many such assemblies, in which everything must be torn apart to get at anything, the Triumph lends itself to easy service. You can work on the engine, or the transmission, or both, but you don't have to split the case and have a bushel of assorted small pieces come raining out on the floor. It is, perhaps, a small point, and the buyer of a new Sports Tiger will go quite a distance before repairs are needed, but when the time comes — as indeed it must for all things mechanical — the relative ease of access to the interior will save the owner money.

The electrical system is one of the AC generator and rectifier combinations that Triumph pioneered. Our test machine was the T100S/R road-rider models, which has a battery, but the T100S/S model — for the competition rider — has direct AC lighting and an energy-transfer ignition system. On battery-equipped bikes, there is a provision for dead-battery starting; a second position on the ignition switch (marked, appropriately enough, "emergency") arranges for the contacts to open when current in the alternator windings is highest. We did not have occasion to use the emergency system, but we understand it gives very nearly the same ease of starting that is true when the normal circuit is in operation — and that is really saying something. The Sports Tiger, just like the Bonneville we tested before, was a One Kick Angel and that kind of behavior makes electric starters look a trifle less attractive.

From the crankshaft, the drive goes through a duplex primary drive to the clutch, and thanks to the unit construction, an automatic chain tensioner replaces all of the adjusting screws that are nearly always present when a separate gear case is used. There is no "lash" in the primary chain and because of this and the rubber-loaded shock-absorbing hub in the clutch center, the drive is unusually smooth and free of snatch.

The transmission is conventional in layout, with constant-mesh gears and a swinging cam plate for gear selection. All of the pieces are sturdy and the whole thing is captured in an exceptionally strong and rigid casing. It is, as not every Triumph transmission in the past has been, strong enough to take enormous abuses without failing. The shifting is slicker than slick and it would take a lot of stomping to bounce the gears into the wrong holes. Neutral, which has to be found before most motorcycles can be started and is more often than not elusive, is a snap to locate on the Triumph. It is a quite distinctive pip up from first gear, or, if you do not trust messages from your toe, there is an indicator needle on top of the transmission that tells the truth and only the truth.

Frame and suspension on the Sports Tiger are like those on all present day Triumphs (and the majority of machines of any make). The frame is tubular, and of the single-loop pattern, with diagonal-braced struts for the rear spring-shock unit mountings and a two-tube cradle holding and protecting the engine's underside. For a bike

that is intended for a wide variety of uses, it is amazingly softly sprung. Naturally, it does not have the true "balloon" ride favored by the more conservative rider, but it is rather spongy by sporting standards. In fact, for anything more vigorous than medium-fast touring, we think the suspension should be stiffened somewhat. We have thought the same thing about some other bikes, too, but they were not propelled along by the Triumph's engine or at the Triumph's pace and that is why we make special mention of this here. The Sports Tiger has a marvelous engine, transmission and brakes, and the rider who wants to use all of these to their potential limit will want to put heavier oil in the front forks, and stiffer springs around the rear dampers.

Generally, riding the Sports Tiger was grand fun, and we were sorry to see it returned to Johnson Motors in Pasadena, who are the distributors here. It was terrifically smooth — which we have already said, but cannot say enough — and it goes down the road like no other 500cc machine we have ridden recently. The clutch action is easy on the take-up, and has a welded-together grip when all the way home. On the Lions Club drag-strip, in Long Beach, California (where we now do all of our acceleration work) we startled all present by going out with lights, mufflers and license-plate, looking very touring on a strip that ordinarily sees only all-out drag machinery, and banged off run after run at over 80 mph. The Tiger was a bit slow away from the line, and that hurt our e.t., but once underway it would really howl. It is not particularly at home on the drag strip, but it certainly managed to rise to the occasion in an honorable fashion.

The Tiger's top speed of 100 mph represents only what it was capable of doing with the standard gearing. The power peak for the engine is reached at 92 mph, and the engine is getting near valve crash when the bike passes the 100-mark. Probably, even though there was no audible rattling, the valves were no longer following the cam at the maximum speed. The flat-out enthusiasts will want to gear this new Tiger differently — but the standard gearing is perfectly selected for all-around riding.

On the whole, we were greatly impressed by the Sports Tiger. It is as nicely finished a bit of machinery as one could want, and the color scheme, in metallic blue and grey, is an eye-catcher if ever one existed. Performance, while not up to the level of the "hot-cam, high-compression" versions one sees winning all of the races, is all that can be had without compromising smoothness and reliability and is pretty good anyway. Finally, is it nice enough and fast enough so that *we* would pay *our* hard-earned money for it? You bet your sweet life it is! •

TRIUMPH T100S/R

SPECIFICATIONS

List Price	$974.00
Frame Type	tubular, single-loop
Suspension, front	telescopic fork
Suspension, rear	spring-arm
Tire size, front	3.25-19
Tire size, rear	3.50-18
Brake lining area, sq. in.	55.2
Engine type	vert. twin, ohv
Bore & stroke	2.72 x 2.60
Displacement, cu. in.	29.8
Displacement, cu. cent.	490
Compression ratio	9:1
Bhp @ rpm	38 @ 7000
Carburetion	1" Amal Monobloc
Ignition	battery & coil
Fuel capacity, gal.	3.5
Oil capacity, pts.	5.0
Oil System	dry sump
Starting system	kick, folding crank

POWER TRANSMISSION

Clutch Type	multi-disc, oil-bath
Primary drive	duplex chain
Final drive	single-row chain
Gear ratio, overall:1	
4th	5.64
3rd	6.70
2nd	9.85
1st	13.7

DIMENSIONS, IN.

Wheelbase	53.5
Saddle height	30.0
Saddle width	10.5
Foot-peg height	11.5
Ground clearance	8.7
Curb weight, lbs.	336

PERFORMANCE

Top speed, average	100
best run	101
Max. safe speed in gears @ 7600 rpm	
3rd	84
2nd	57
1st	41
Fuel consumption range, mpg	55/70
Mph per 1000 rpm, top gear	13.2

SPEEDOMETER ERROR

30 mph, actual	29.8
50	49.2
70	68.9

ACCELERATION

0-30 mph, sec.	2.5
0-40	3.8
0-50	5.7
0-60	8.0
0-70	10.6
0-80	14.0
0-90	19.6
0-100	31.0
Standing ¼ mile	15.8
speed reached	84

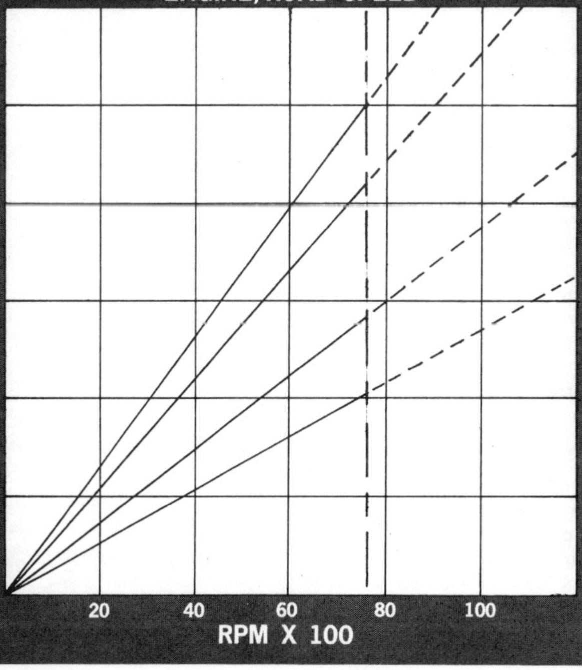

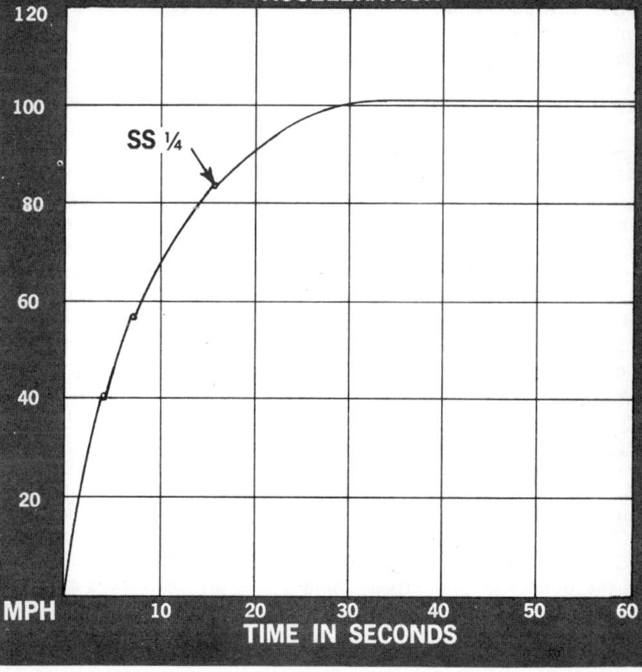

TRIUMPH FOR 1963

500cc TIGER 100

650cc TROPHY TR-6

650cc THUNDERBIRD

650cc BONNEVILLE

200cc TIGER CUB

200cc SPORTS TIGER CUB

Triumph distributors have introduced a ten-model motorcycle program for 1963 with the introduction of a new automatic scooter. Emphasis has been placed on "cleaness" (not our spelling, Ed.) of design and engineering improvements. "Prestige" model of the line is the T-120 Bonneville powered by a new engine with special dual carburetor cylinder head. Also, the new 650 models are claimed to be smoother at all speeds owing to a new engine balance factor, new frame mounting and rubber cushioned handlebars. There are five 650cc models, all with the new engine, including two Bonnevilles, one for the road rider, the other with specifications designed for the competition minded.

Two Trophy Sports models, one for highway use, and another for the sports minded, are identical to the Bonneville but employ a single carburetor head. The 650cc Thunderbird is designed for street use and while powered with the same engine as the Bonneville and Trophy models the engine uses milder 7 to 1 compression pistons and milder camshafts. Three models in the 500cc category are introduced for '63, two sports models and one on the order of the Thunderbird. Two Sports models of the Tiger 100 series are identical and are fitted with the new unit construction engine. One, designated the T-100-SR, is equipped with battery ignition; the other, T-100-SC, with magneto.

The 500cc Speed Twin is a highway model with a mildly tuned engine and offers low maintenance and good performance. Two 200cc ohv single-cylinder models form the Triumph lightweight series, the Tiger Cub T-20 for practical transportation, and the T-20S for the competition minded, both employing the new unit-construction engine. The Sports Tiger Cub is fitted with a high performance engine, large wheels, smaller gas tank and sports fenders.

The scooter division offers the new Tina with an impressive list of assets, not the least of which is its automatic operation with only two controls, called Tina-Matic. Triumph motorcycles, parts and accessories are distributed in the U.S. by the Triumph Corp., Towson, Baltimore, Md., for 39 Eastern States, Johnson Mtrs. Inc., Pasadena Calif., for the 19 Western States. •

CYCLE WORLD ROAD TEST

TRIUMPH T-120 TT SPECIAL

LAST AUGUST, out on the Bonneville salt flats, a Triumph powered, streamlined motorcycle made some runs that resulted in impressive new entries in the record books: two-way averages of 230.3 mph under the free-fuel rules, and 205.8 mph on gasoline. Of course, this is old news to motorcyclists, as the event was reported by the motorcycling press at the time; what is not generally known is that the records were set with a pair of engines that did not differ, except in detail, from those used to power the stock Triumph Bonneville sports/touring bike. And, all of the special parts used on those two engines are available to anyone from the parts bins at Johnson Motors, in Pasadena, California, who are the Triumph distributors for the western United States.

The record engines were prepared in the Johnson Motors shops, under the direction of Pete Colman, and all of the work was done in only two days — which will give the reader an idea of what a straightforward job it really was. Bonneville engines were used, and the only external modifications were the addition of 1½-inch Amal GP carburetors; the installation of which involved boring out the ports to a huge oversize and welding-in stub pipes of the appropriate diameter. Short lengths of radiator hose were used to connect the carburetors and their mounting-flange stubs to the stubs on the cylinder head, and these served not only to connect the manifolding, but to act as vibration and heat barriers as well. Actually, when the engines were delivered to Joe Dudek, who built the streamliner, the 1½-inch carburetors were too large to fit inside the shell, and 1¼-inch Amal GP carburetors were substituted — which makes the final results all the more impressive. This should give anyone who might be shooting at the record something to think about: Dudek can always make room for the larger carburetors, call Bill Johnson, who does the rider's chores, and go back to Bonneville with excellent prospects of going even faster.

Inside the record engines, there are a few additional modifications. The crankshaft's ball-type main bearings were replaced by selected-fit, single-lip roller bearings and the lower-end assembly was carefully balanced. High compression (11:1) Robbins pistons were fitted. The camshafts used were Jomo (for Johnson Motors) #15,

Bennie Sims turns it on for a 100 mph run.

which in conjunction with Triumph's E4040 tappets, give valve timings of 41° BTC – 71° ABC for the intakes and 71° BBC – 41° ATC for the exhausts. The intake valves were Jomo's 1⅝-inch racing replacements, and these were installed on seats that had been trimmed to give the greatest effective diameter. Other valve-gear modifications were the polishing and lightening of the valve rockers, the use of lightweight pushrods (from the "C" range engines) and the installation of Jomo progressive-rate valve springs.

The only modification made in the interest of reliability (which the Triumph has, in full measure, as it comes from the factory) was the addition of a tube that carries oil from the pressure-regulator bypass back to the oil tank. In the stock engine, this oil is dumped into the engine sump and picked up by the scavenging pump, and that arrangement works very well. However, at sustained high engine speeds, when a lot of oil is being bypassed (due to the great quantities of oil being pumped), the sump tends to load a little, and the re-routing of bypassed oil eases the task of the sump-scavenging pump.

A true measure of the reliability of this power unit is the fact that it did its job with a minimum of fuss and a maximum of reliability. Over 17 runs were made down the 9-mile stretch of salt at speeds well over 200 mph. Really though, this is little more than could have reasonably been expected. The same engine is being used for TT racing all over the country, and Triumphs win with monotonous regularity – and you can't do that unless you finish a race with a healthy engine.

A newer, and slightly improved, version of the engines used to set the record powers the 1963 Triumph Bonneville special – the machine that is the main subject of this report. It follows the long-time Triumph layout, but has a unit-constructed crank and transmission casing – and other detail changes.

Bore and stroke dimensions remain unchanged, but the crankshaft and flywheel, and the main bearing layout, have been improved. Previously, the crank was carried in one ball (on the drive side) and one plain bearing, but heavy-duty ball bearings are now used on both sides. "H" section, aluminum-alloy connecting rods are again featured, and these have shell-type, replaceable insert bearings. The flywheel is new, as is the method of attaching; the balance factor has been increased to 85-percent (of the reciprocating mass). This, with changes in the frame we shall cover later, makes the new engine the smoothest of all the big vertical twins.

Modifications have been made in the engine's upper end, too. The barrels are of long-wearing alloy iron, and carry a revised cylinder head stud pattern, with an extra stud tucked in between the bores to help the 8 that were there before. An extra fine touch is seen in the groove machined around the top of each bore on the upper surface of the barrels. The gasket bulges down into this groove when the head is torqued down and it grips the gasket so as to prevent it from blowing outward.

Triumph valve gear has always been unexcelled and it is continued unchanged, but there is a new cylinder head, which has a greater cooling-fin area. Also, the valve-rocker boxes have had finning added; partly in the interest of appearance, but also to give better cooling and rigidity. Material has been added at strategic points in the head casting that, in combination with the extra stud between bores, gives extra rigidity there, too, and holds valve seat distortion to a minimum under conditions of high temperatures and high gas loadings. In anticipation of the sustained high speeds at which these engines are to run, the cam followers have been given brazed-on face blocks of some wear-resistant material – possibly steelite or a near equivalent.

The camshaft drive gears have been widened, for

Engine of the Dudek/Johnson Bonneville streamliner, the world's fastest motorcycle. All parts used in the engine to achieve the 224 mph International record are available as accessories.

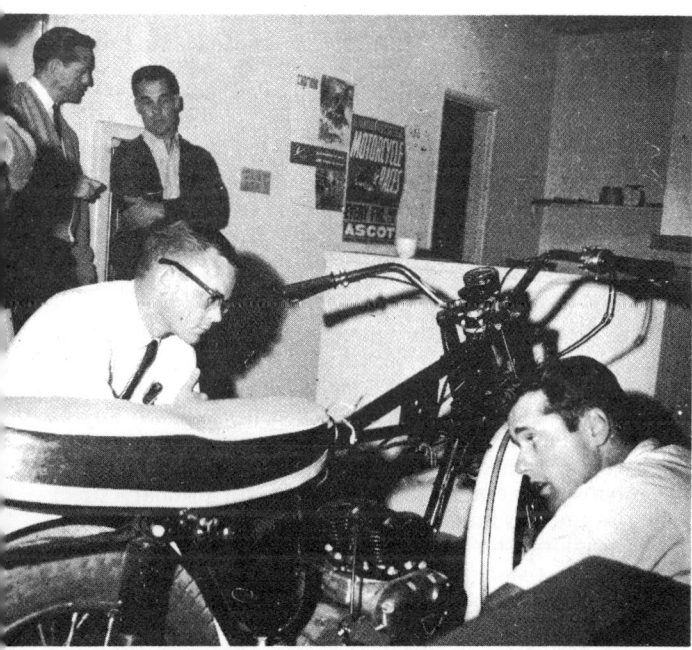

CYCLE WORLD'S Gordon Jennings and Pat Owens, Johnson Motors Service Dept., discussing the T.T. Special's engine. In the background are Johnson Motors' Don Brown and Triumph racer Skip VanLeeuwen.

more quiet running and added reliability, and an oil-slinger has been incorporated next to the oil breather to eliminate oil loss at that point. A tachometer drive is provided on the left end of the exhaust camshaft and all Bonneville models carry "N" camshafts on the exhaust side; "Q" camshafts are fitted on the intake side. This gives a fairly sporting intake timing combined with a relatively mild exhaust timing, with the result that the engine has a lot of power over a wide speed range — and will idle down nicely, too. Of course, both Bonneville camshafts are a bit more sporty than those found in other engines in the "B" range.

A new, and very worthwhile, change for 1963 is the revised ignition system. There is no distributor, as such; a point plate is housed in the timing case cover, and the breaker cam is driven by the exhaust camshaft. Two points are used, and each of these controls what is, in essence, two entirely separate ignition systems: there are pairs of coils and condensers — one for each spark plug. With such a direct drive, the ignition timing is very precisely controlled, and the system has the added advantage of being compact and light.

Actually, the Bonneville is available in two stages of tune: the first is the Bonneville "Speed Master," which has 8.5:1 (compression ratio) pistons and dual 1 1/16-inch Amal Monobloc carburetors. However, here in the west, Johnson Motors is offering a model called the Bonneville T-120 TT Special, which the Triumph Factory has kindly consented to produce for them, to their order. This machine, an example of which was loaned to us for this test, is delivered without lighting equipment and has 12:1 (egad!) pistons and 1 3/16-inch carburetors. Where the standard model has 50 bhp at 6500, the TT Special is said to have 52 bhp. Actually, we think the gap between the pair is greater than that; or, perhaps, the peaks given are the same but the TT Special has a lot more power over a wide range of engine speeds on both sides of the peak.

As though to confound our technical editor, who was holding-forth only last month on the great virtues of double down-tube frames, Triumph has abandoned that arrangement in favor of one having a single, large-diameter down-tube, which hooks into a two-tube cradle under the engine. Although this might seem a retrograde step, even the technical editor will admit that the results justify the change. No alteration has been made in steering angle, or trail, or wheelbase, but the new Bonneville feels more stable, and solid, and the frame tends to subdue engine vibration to an amazing degree. While it is not widely known or appreciated, at least half of the battle against vibration in a motorcycle is fought in the engine mounting and frame. When the engine, its mountings and the frame are right, the rider will not feel much vibration; when they are wrong, he will get a real vibro-massage. In the case of the Bonneville, we are of the opinion that last year's small-diameter down tubes, even though structurally correct, would "twang," just like a pair of fiddle strings, and the rider was painfully aware of this at times (those times being when the frequency of engine vibration matched the natural frequency of the frame tubes). The big single-tube now used is probably little, if any, stronger, but it will not buzz, as the two lighter tubes did.

Vibration damping has been provided in the drive system, too. A duplex chain, with a blade-type tensioner, takes the drive to the clutch — which has a hub that carries a 3-vane shock damper. There is no snatch or pounding in this system. The clutch, incidentally, is new, with the plates driven by a cast-iron housing. The friction material is bonded to the clutch plate and the clutch grips fantastically well, while providing what is surely

19

the smoothest engagement to be found on any big-inch bike. Part of this is due to a redesigned clutch throwout, which now has a wedge-ball device to force the clutch-rod through and push the plates out of engagement. The device in question provides a high mechanical advantage, for light lever pressure, with a minimum of friction.

Those models in the "B" range sold primarily for competition have special electrical systems. Current for the ignition is supplied by a rotating-magnet AC generator, mounted at the drive-end of the crankshaft. This unit will feed enough current even at kick-over rotating speeds to fire the engine, and no battery is required. And, of course, once the engine is running there is more than enough "juice" to make the sparks for high-speed running. Certainly, one could not ask for better results than we got from this system during our test. The engine would start easily when cold — even though fitted with Lodge R49 racing spark plugs. And, obviously, that 12:1 compression ratio, with its attendant high cylinder pressures, does not make the ignition's task any easier, but the system performed flawlessly nonetheless.

We took the Bonneville TT Special to Harry Schooler's Ascot Park to photograph it in its natural habitat. From the pictures it can be seen that the track was in the arduous, expensive and time-consuming stages of preparation for the Jimmy Phillips Memorial 100-lap TT which followed two days later, and was won by Skip Van Leeuwen riding a similar machine on its first outing.

Performance-testing of the Bonneville TT Special was done at two different sites: the Long Beach Lions Club drag strip; and for handling evaluation and high-speed testing, Riverside Raceway. The machine was in virtually "from the crate" condition, except for the substitution of downswept exhaust pipes with reverse-cone megaphone. The gearing was, for the high-speed portion of the test, the standard 4.84:1. Later, for an all-out attempt at the drag strip, we installed a 54-tooth overlay sprocket on the rear wheel, changing the gearing to 5.68:1. Pump gasoline was used and no oil was lost or added during the entire duration of testing.

While at Riverside Raceway we discovered that the TT Special was not very fussy about its mixture. Our first run, on 330 main-jets, was at exactly 100 mph. Going up through the jets in gradual stages nudged the speed upward until we got up to 380 jets, and we finally settled on 370 jets as being best for that day. In all, we made what seemed like two-dozen flat-out runs, increasing speed with each try, until we finally got a best run of 123.5 mph — and that was done under the handicap of a rather bothersome side wind. Five consecutive runs were made at speeds over 120 mph. The limit, when it was finally reached, was actually set by the amount that the engine would over-rev without losing too much power, which was about 7900 rpm (the tachometer flickers, even the rider's eye-balls flicker, at that speed and it was impossible to get a really accurate reading). With one more tooth on the countershaft sprocket, we think the stock Bonneville TT Special would push up to about 127 mph, and that is fearfully fast for a machine that can be bought in that condition right from a dealer's floor.

Handling was, as we have said, better than we remembered of the 1962 Bonneville. It was very stable, and yet responsive, and we would someday like to try one with proper road-racing tires and high pegs. While making those high-speed runs in a cross-wind, we discovered that the high-speed response was especially good. We were following a white line marking the center of the straightaway, and this gave us a point of reference that allowed us to observe that we were running banked over into the breeze. Interestingly, when the wind was gusty, the bike would tilt sharply into the gusts and seemingly adjust itself to the side pressures without conscious effort from the rider and without straying from the line.

The only complaints we have are that the steering damper had a tendency to vibrate loose, whereupon the handling would become a trifle twitchy, and that the brakes were not quite up to the speed potential of the machine. When the brakes were applied hard at the end of a high-speed run (there is a turn at the end of that Riverside Raceway straight), they would shudder noticeably. The bike would pull down quickly enough, but we got the distinct impression that there was little, if any, braking capacity in reserve. Perhaps it is unreasonable to ask for perfect braking on so rapid a mass-produced motorcycle.

After completing our high-speed tests, the TT Special was hauled off to the drag strip and, without changing the gearing, given the grand old American ¼-mile "banzai" try. Pulling the tall gearing, and carrying our 190 lb. (in leathers, helmet and boots) regular test rider, the bike immediately cracked-off a 14-second, 93.4 mph quarter. Then, to see what the TT Special would do if given a chance, we changed to the 5.68:1 gearing and brought our good friend Bennie Sims (a professional Expert-class racer, and husband of Carol Sims, CW's Associate Editor, who weighs about 140 lb. ready to go) into action. Under Bennie's able throttle-twisting and handlebar pointing, the bike soon gave us an even 100 mph at the end of the quarter, with an elapsed time of only 13.34-seconds.

As a final measure, just to assure us that we had been running the Real Thing, and not a super, super-tuned Triumph, the bike's engine was torn down for our inspection, and duly certified as stock. We are also happy to certify that despite the unmerciful flogging the bike received, the engine was still in excellent condition.

Actually, the stockness of the TT Special is of limited importance, as the machine will be delivered with any of several engine and drive options, at the customer's request. Even so, we think it is interesting, and there is the fact that many people are buying this bike for ordinary street riding — and it is mild enough for that sort of running, too. While primarily for racing, it can be outfitted with lights, and it has, as standard, Triumph's big, comfortable touring-type saddle. And, there is no disputing that it is, among the machines we have tested, the fastest of them all — regardless of displacement. The performance-oriented rider, whether he intends to race or just likes a lot of power on tap for touring, will be hard put to find more sheer flashing speed than is provided by the Bonneville TT Special. •

TRIUMPH BONNEVILLE TT SPECIAL
SPECIFICATIONS

List Price	$1,158, F.O.B., P.O.E.
Frame Type	tubular, single-loop
Suspension, front	telescopic fork
Suspension, rear	swing arm
Tire size, front	3.50-19
Tire size, rear	4.00-18
Brake lining area, sq. in.	32.5
Engine type	vert. twin, ohv
Bore & stroke	2.79 x 3.23
Displacement, cu. in.	39.6
Displacement, cu. cent.	649
Compression ratio	12.0:1
Bhp @ rpm	52 @ 6500
Carburetion	(2) 1 3/16" Amal Monobloc
Ignition	alternator and coils
Fuel capacity, gal.	3.75
Oil capacity, pts.	6.0
Oil System	dry sump
Starting system	kick, folding crank

POWER TRANSMISSION

Clutch Type	multi-disc, oil-bath
Primary drive	duplex chain
Final drive	single-row chain

Gear ratio, overall:1
4th	4.84
3rd	5.76
2nd	8.17
1st	11.8

DIMENSIONS, IN.

Wheelbase	55.5
Saddle height	31.0
Saddle width	10.5
Foot-peg height	11.0
Ground clearance	8.0
Curb weight, lbs.	390

PERFORMANCE

Top speed	123.5

Max speed in gears @ 7900 rpm
4th	123
3rd	105
2nd	74
1st	51
Mph per 1000 rpm, top gear	15.8

SPEEDOMETER ERROR
30 mph, actual	no speedometer
50	
70	

ACCELERATION
0-30 mph, sec.	2.4
0-40	2.9
0-50	3.5
0-60	5.5
0-70	7.3
0-80	10.2
0-90	13.1
0-100	16.6
Standing 1/4 mile	14.0
speed reached	93

CW Trail Test TRIUMPH TRIALS

TRAIL TESTING TAKES A NEW TURN this month with the familiar Triumph 200 Cub in trials form, a machine as near ideally suited for trailing as it can be. Regular CW readers will recognize our test bike as the same as seen in last month's "Trials, Anyone?" feature, and the Editor's transportation when covering the 200 mile National enduro in San Luis Obispo, California, in our May 1963 issue.

We may have given enough justly due praise of this machine in the aforementioned feature; we did, after all, detail quite thoroughly its perfect suitability as a trials machine due to its low gearing, wide torque characteristics, well designed steering geometry and suspension, etc.

In truth, we disliked very little on this unique motorcycle. Its professional prowess as a championship trials bike is constantly being re-proven in England where it reigns as one of the most successful and popular bikes in this very demanding sport. In scrambles form the Tiger Cub, or TS-20, is well known on both sides of the Atlantic; as a trials machine, though, little has been demonstrated over here. Though it is available in extremely small quantities, the T/20-R differs from its hotter cousin primarily in its gearbox, a widespaced affair, and basic gearing. The Tiger Cub, or scrambler, can easily be geared to equal the Trials model by use of a high overlay rear sprocket. Only one feature on the trials model cannot be duplicated on the scrambler, the trials gearbox employing two extremely low first and second gears, and the two "normal" third and fourth gears. This allows trailing in the worst country with ease, and near-highway cruising where allowed. Loss of this feature when converting the Special Tiger Cub to Trials specifications is a minor detriment and should not bother the trailing rider.

Lightness, a virtue without compromise when trailing, is an exemplary feature on the 223-pound Cub; the lightest or least muscularly endowed rider can easily master it. Low horsepower, only 10 for the 200cc (12.2 cubic inch) single-cylinder, overhead valve engine, though seemingly a drawback, is used to an advantage by allowing extremely slow engine revolutions thereby permitting slow riding without the usual stalling and excessive clutch manipulating.

Though the 16 horsepower accredited to the other Triumph Cubs does not offer quite the handiness of the Trials, it should not be too much of a shortcoming with proper gearing. The almost all-alloy wet sump engine is cradled in a single down-tube tubular frame. It is an in-unit with gearbox power plant and being a motor-cycle, is controlled entirely as one with a neat, short positive acting foot shift controlling the four-speed gearbox. Alloy is used lavishly all about, accounting further for its light weight, in such vital places as the fenders and huge rear sprocket.

Handlebars are in the traditional European trials manner, wide spread and decidedly flat. We would recommend a higher bar if extensive trailing was contemplated. Ball-end clutch and brake levers are standard, as is a silly little bulb horn which qualifies it as a true trials model in Europe. Rubber covers adorn the sliding fork joints to keep out dust, a luxury feature but useful. Instrumentation is limited to a speedometer, the use of which is sparse on such a bike. •

WE TRIUMPH AT BONNEVILLE

Portrait of a record setter, CYCLE WORLD style.

Handsome Sonic fairing contributed considerably to success.

LAST SPRING, when we tested the very rapid Triumph Bonneville TT Special, gearing and a shortage of straightaway room prevented us from determining the bike's absolute top speed. We were continually running out of engine revolutions and space, and even so the TT Special pushed up to 123.5 mph — fast enough to satisfy nearly anyone; yet not the absolute limit of the bike's potential. Since that time, there have been endless discussions around these offices concerning the Bonneville TT Special's *real* top speed. Inevitably, someone (our technical editor) proposed that we take one of these machines to Bonneville. Our plans for the magazine included coverage of this important speed event, and it seemed a simple matter to haul along a bike; a little something to occupy our time while not taking pictures or gathering information. Interest in the project sharpened when it was pointed out that there was a record in the 40 cubic inch, class-C, partially streamlined category that stood at just 127.774 mph — a speed obviously within the capabilities of the TT Special.

Once the decision had been made (by the tech. ed.) to try for the record, and the financial sponsorship found (our publisher is generous to a fault), there remained only the job of preparing the bike. We knew that preparation was going to be a snap; it had to be: we did not decide to go until just one month before the Speed Trials began.

With a whole month to squander on the project (and it was all to be leisure-hours work) we obtained the essentials: a Triumph Bonneville TT Special; a Sonic road-racing type fiberglass fairing; and sundry bits and pieces from Johnson Motors' stock of Jomo accessories — these will be itemized and explained later.

The first job was to strip the motorcycle in preparation for some rather vital modifications that had to be made for the special task it had to perform. Then, with innards strewn all over the garage, work was started on the engine.

At sea level, or anything near it, the stock engine would have provided all of the power needed; unfortunately, the Bonneville salt flats are up at 4300 feet, and the air is so thin at that altitude that engines tend to get short of breath, as do people. To compensate for the power loss, 1/8-inch oversized intake valves we installed, along with Jomo #15 camshafts and racing tappets. No changes were made in compression ratio — we retained the stock pistons, trimmed a bit to accommodate the bigger valves — and the only other engine modifications were in the lubrication system.

In the stock Triumph engine, oil released from the pressure-limiting bypass is drained into the sump, where it is picked up with return oil from the bearings, etc., and pumped back to the oil tank. This system works very well unless the engine is operated at very high rates of revolution for extended periods (which was precisely what we had in mind); then, the flow into the sump is sometimes too much for the scavenging pump to handle, and the sump tends to load with oil. To insure that this did not occur, we installed a fitting in the end of the bypass valve housing, in place of the in-

Special rear wheel allowing use of small sprockets was used. Note special 200 mph tire and straight exhaust pipe.

Stock 19" front wheel was replaced with special brakeless 21" wheel for record attempt.

Johnson Motor's Don Brown congratulates Jennings on his record run.

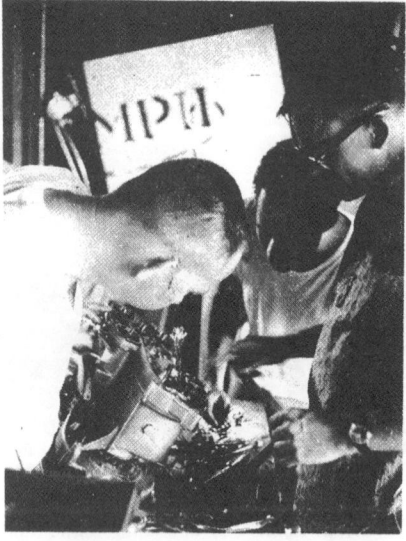

Now how do you suppose **that** comes out?

Jennings at work early in the game; one week to go!

"Rich" Richards changes the countershaft sprocket as Jennings looks on.

Tech. Ed Jennings awaits the word to go with Wayne Moulton standing by.

BONNEVILLE TRIUMPH

dicator rod, and lead the bypassed oil directly back to the oil tank. This was the extent of engine modifications; we retained the standard 1 3/16-inch Amal Monobloc carburetors, and even used the exhaust pipes supplied with the a machine: 1 3/8-inch upswept high-pipes of the type intended for scrambles and TT racing. No megaphones, no anything; we wanted to remain as near "stock" as is possible.

A lot of the "stock" approach went right out the window when we got to the matter of tires and gearing. It is required by regulation, at Bonneville, that high-speed machines use high-speed tires, and while such tires were available for the TT Special's 19-inch front wheel, the 18-inch rear wheel was no-go. Moreover, it is not possible, within the dimensions of the Triumph's cases, etc., to get the kind of gearing that would allow the kind of speeds we had in mind. Thus, it became necessary to fit a special rear wheel, which had the brake drum moved over opposite the sprocket. This permitted the use of substantially smaller than standard rear wheel sprockets. Luck was with us in this regard, for our good friend, Rich Richards (he of the potent drag-strip and Bonneville, 40-inch fuelers), had just such a special wheel — complete with a 20-inch rim and approved, high-speed tire. We were now all set in the wheels and tires department; or so we thought.

At this point we had the engine back in the frame, and the wheels mounted, and all that was required to make the motorcycle function was to rig a crossover linkage for the brake lever, which was on the left; the brake drum being now located on the right side of the wheel. The mounting plates on each side of the frame, which support the back of the transmission, were bored through, and a cross-shaft installed. A couple of levers and linking rods completed the connection to the brake and here again we fancied that we had the situation gripped tightly by the hindside. A hurried reassembly after a sprocket change was to produce errors that made this estimate of the situation appear, in retrospect, somewhat optimistic.

The fairing was fitted — a less difficult task than one might imagine — and sent out to be painted and we then faced the problem of a seat and fuel tank. The stock components would not do: the tank was wide enough so that the low, "clip-on" handlebars we had installed made contact even at small steering angles, and the seat's deep padding, while superbly comfortable, raised the rider's rump upward in a most unaerodynamic fashion. Again, it was Jomo to the rescue, with a racing seat and fuel tank much better suited to the job. Both were of molded fiberglass, and the seat entirely covered with leatherette. These are new items in the Jomo line, made to their specifications by Custom Plastics — who do some of the best work in fiberglass we have ever seen.

All of this takes a trifle longer than the telling, and by the time the bike was ready, we had to work right down to the last moment and went flying off to the salt flats without ever having even started the TT Special's engine. Our naturally high hopes were leavened by the nagging knowledge that everything was not as it should have been: there was a suspicious tightness in the engine when the pistons were eased past top-center. However, time had run out and it was time to leave. Part of the crew had a job of reporting to do and there was no time to dawdle.

That suspicious tightness was corrected after our arrival at Bonneville. Rich Richards, who is more than a little familiar with Triumph engines, offered the opinion that the pistons were contacting the intake valves. So sure of his diagnosis was he, that he pitched right in and helped us pull down the upper end of the engine for a look. He was disgustingly correct: in the great rush to do it all in one month, the necessary check of clearances had not been made and the intake valve heads were just pipping the pistons.

To correct this, the valves were pocketed into the head a bit deeper, the engine reassembled, and a (by this time) weary CYCLE WORLD crew wheeled their entry out to the salt. We were greeted with polite but unimpressed interest by the Bonneville regulars, who eyed our straight pipes and "showroom" carburetors dubiously, and whose faces clearly showed that they considered us mere babes in the woods. To tell the truth, after looking around at the exotic and highly modified machinery there (you have never seen so many GP carburetors in one place), we were beginning to feel a bit the same way ourselves.

The fretting and doubt was for nothing. In what remained of the day, we tuned up to within 1 mph of the record, and went back out early the next day with fire in our eyes (and stomachs, too: we had celebrated that night). The first order of business was a change of gearing, promptly done, and then we made a run at just over the existing record, which qualified us to make an official attempt the following morning, and then capped it all with another run (after a change of jetting) at 135.74 mph. Ah the joy! To sit in the shade and watch the scoffers edge up to take another look and walk away muttering to themselves.

To those readers who don't think 135 mph is fast, we would like to say that under ideal conditions it might not be; but the salt was very rough and slippery this year, and the barometric pressure was going up and down like a yo-yo, upsetting everyone's carburetion, and at that point in the speed trials, we were going faster than almost everyone. Faster, in fact, than anything else in the 40-inch class, and that included some fast nitroburners. Also, we were faster than all but a couple of the big V-twins. •

PHOTOS BY CAL WEST

Brilliant chromework is featured throughout Catuto's '57 Triumph, yet manages to avoid an overdone look so prevalent in many customs. Entire frame is done in triple plate; oil tank, all brackets and braces, nuts, bolts, and fittings were dipped. Engine parts of aluminum were chromed rather than buffed, and even carb made a trip to the plater's. Two-and-one-half years have been expended by the 28-year-old Angeleno, as well as about $3000. Strikingly different is the asymmetrical paint treatment on the tank and rear fender, done by the skilled hand of Slimbo, Redondo Beach lacquer lad. Maroons, silvers, purples and golds are mated in surprising harmony. Other lauds for construction assists go to Big City Sales and the Motor-Cycle Shop, also numbering among the more sensible, progressive dealers.

Something for a STARTER

A strong appreciation for functional as well as beautiful machinery is a natural by-product of Richard Catuto's favorite job: part-time starter at Los Angeles' famed Ascot track. Dick has been handling the flag-waving chores during cycle racing events for several years, and with some twenty previous bikes during the past 14 years, has developed some solid ideas of what his ideal street scooter would be like. Performance was a must — mere paint and chrome would not satisfy the requirements. Sensible handlebars to allow comfortable, safe handling under all conditions were selected, similar to the handles on the best TT and oval-trackers which face Dick on those action-packed outings. Whipping wheelies at will, none can deny that Dick's ability and TR-6 power teams up well.

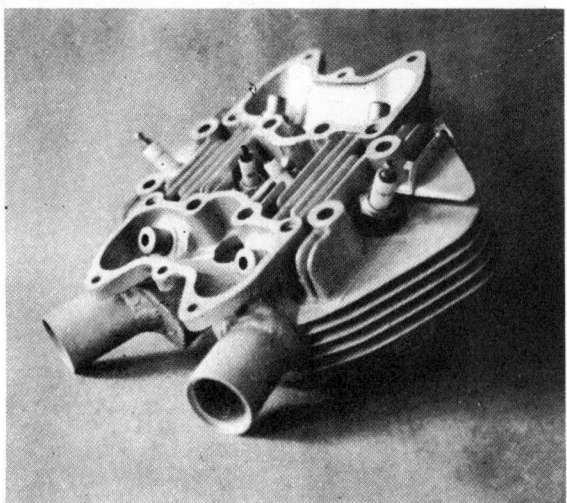

The T-100 cylinder head has welded-in inlet stubs, and fairly bristles with 10mm and 14mm NGK spark plugs.

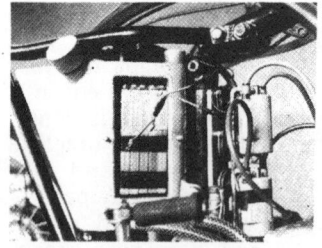

Battery is housed in the special oil tank; the two-lead coils get low-tension current from this source.

H&C

Outstanding performances by Triumph motorcycles in our TT races are nothing new, but recently the crowd at Ascot Park witnessing the 100-lap TT was treated to a performance by one such Triumph that they will never forget. Clark White, riding a 650cc Triumph twin, moved up through the field (which included some exceptionally good bike/rider combinations) to take the lead on the 3rd lap and was never in any danger of losing it from that point to the finish. It was, obviously, a fine exhibition of rider skill, but anyone who saw the race must surely have noticed all the help White was getting from the machine. His Triumph was clearly faster than anything else there, and considering the length of the race, it was clear that the power had not been gained at any cost to reliability. We decided, then and there, to make it our business to find out more about this particular TR-6.

The machine is owned and tuned by Gary Robison, of Harman and Collins Engineering, and he very kindly consented to lay bare its internals for photos, and to tell us all that had gone into the modification and preparation of this Triumph. The modifications are, as you will see, quite extensive, and include not only the engine, but the brakes and the oil tank.

Starting at the top of the engine, we find that the cylinder head has been subjected to a fairly radical reworking. The head casting is from an old Triumph T100 twin, with special stubs welded into bored-out intake ports. There is now an extra two-degrees of "downdraft" in the port shape, and they are splayed out a bit more than stock. Inside, the ports hold a constant 1-3/16" diameter right down to the valve guide, where they begin to open, reaching maximum diameter where it rolls out into the valve seat.

The intake valves are reworked Chrysler parts, 1-5/8" in diameter, with a pronounced "tulip" shape that Robison says improves airflow past the valve head considerably. These valves, which have smaller stems and are slightly lighter than the stock Triumph parts, are fitted in Matchless guides. The exhaust valves are 1-3/8" James oversize replacement parts.

Because the Triumph engine has a true

Here, peeking over the primary drive case, can be seen the cover on the special ignition distributor.

hemispherical combustion chamber, the chamber is quite deep, and to get a racing compression ratio, it is usually necessary to employ pistons with very high domes. This gets the job done, obviously, but it has some disadvantages. With the piston crown up in the head, the combustion chamber shape is roughly like an orange-peel, with a long flame travel into the far corners and no "squish" effect. Also, the high dome tends to partially mask the valves, and has something less than a beneficial effect on breathing. To overcome these problems, the combustion chambers were reshaped with a deep recess all around the outer rim. Then, special pistons were made that extend well up out of the cylinder and into the recess rimming the combustion chamber, but have a relatively low crown. Through this modification, Robison was able to reduce flame travel in the combustion chamber and leave the valves quite clear. Also, as the area around the pistons comes up to within a few thousandths of the recess machined in the edges of the combustion chamber, a useful squish effect is

Piston nearest is a special part, with a lowered dome and edges raised to fit in recess machined in head.

produced, and that is a great aid to combustion.

The compression ratio used with this engine is 10.5:1, which is lower than the absolute limit the engine will tolerate, but at that ratio the engine runs cool and very smoothly, and there have been no problems with detonation occurring under full-throttle, low-speed conditions, which is often encountered in deep-breathing modified Triumphs having a higher compression ratio. CYCLE WORLD's record-breaking Bonneville was plagued with this problem. The engine would detonate heavily if full-throttle were applied at anything under 6500 rpm, after which volumetric efficiency began to fall and the thermal loading was not so high.

Further combustion smoothness in the Harman and Collins Triumph was obtained after making a really radical change in the ignition system. An extra spark plug was fitted in each combustion chamber, across from the existing plug, and as this fires the mixture at two widely separated points, flame travel is yet further reduced. Due to a shortage of space, and because the extra plugs lean toward each other in the center of the head, 10mm plugs were used in the additional plug holes, although Robison retained the 14mm plugs in the original holes. NGK spark plugs are used all around, and Robison is enthusiastic about their performance.

To fire these four spark plugs, an entirely new ignition system has been developed. In place of a magneto, or energy-transfer system, a special distributor was fabricated, to mount on the back of the timing case in the spot normally occupied by the magneto. This distributor contains a Honda point plate, and points (there is no advance mechanism) and the breaker cam has two lobes, so that both sets of points open and close together. Now, if you are thinking that this would fire both sets of plugs at the same time, you are quite right. Each set of points controls a Honda 6-volt, double outlet coil, which is wired so that the high-tension current path goes out one lead to a plug, across to ground, and then from ground back through the other plug to the second lead and back to the coil, completing the cir-

peak but with more through the lower engine speed range, is obtained with 42-degrees of advance. This suggests that perhaps in splaying the ports wider, so much intake swirl has been induced that it begins to adversely influence combustion at high engine speeds unless the ignition advance is reduced. Of course, the effects of intake swirl are much more pronounced at high speeds, and that would explain why the 42-degrees of advance gives more power at lower speeds, where swirl effects are reduced.

With Gary Robison's firm, Harman and Collins, sponsoring this Triumph, it is no more than logical to expect that the machine would have H&C cams. These are the H&C #9, of a newly developed series, which have an extended duration and .015" less lift. These cams also have "non-harmonic" ramps, which do not excite chattering in the valve gear and which hold off valve float until a higher speed than was once thought possible. Ordinary radiused slipper-type Triumph followers are used in conjunction with these cams and although this particular engine has

torque appears to be improved by the long inlet pipes, but their length was determined more by convenience than science. The exhaust pipes are very large in diameter, 1-7/8", and again, the length was determined by convenience. To get an unrestricted flow into these big pipes, Robison removed the steel exhaust stubs from the Triumph head, and bored the outer portion of the exhaust ports large enough so that the pipes spigot in. The pipes are retained by springs, stretched between the head and clips welded onto the tops of the pipes near the head.

A more or less stock lower end assembly is used. The balance factor has been altered, but the flywheel, bearings, etc. are quite stock. There is, however, one major modification at the engine's lower end: instead of having a "timed" breather, which forces the crankcase to function as a pump, Robison has grafted onto the timing case a Parilla baffled breather that vents directly to atmosphere. From the Parilla breather, a large diameter plastic hose leads upward and back to a point above the rear wheel. Additional, smaller breather hoses lead back from the valve chests on top of the cylinder head. Experiments have shown a slight, but very clear gain in power with the revised breather system, and it offers the additional advantage of reducing oil consumption (actually, oil loss) in a long race.

The oil supply is carried in a special, light-alloy tank that does triple service. First, it holds the oil, and holds it in out of the way. Second, because of its location, it serves as a sort of splash guard, or fender, for the rear wheel. Finally, it holds the battery for the ignition system in a sealed and rubber-padded compartment. This battery is a special, lightweight item borrowed from the aircraft industry, and it is made up of 5 cells, each with a separate no-leak plastic case.

The rest of the bike is a relatively standard 1962 Triumph Bonneville. However, the front brake is one of Albert Gunter's disc units, with an Airheart caliper. This brake is actuated through a "Go-Kart" master cylinder (also made by Airheart) mounted on the handlebars and linked to the brake lever via a short rod. Clark White is extremely impressed with this brake. The standard rear brake faded away completely before the 100 lap TT was half finished, but White found that he could continue charging hard into the corners using the front brake alone, and it never gave any indication of weakening.

TRIUMPH SPECIAL

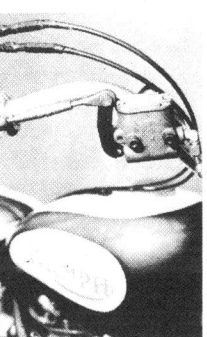

Albert Gunter's special hub and brake disc, with Airheart caliper, gave really exceptional stopping power.

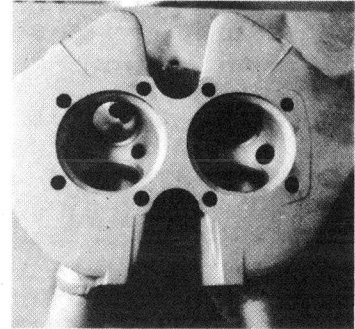

Combustion chambers are remachined to provide squish area.

cuit. It would be possible to use a normal firing arrangement, with one set of points being actuated and then the other, but in the setup chosen each coil has leads to both cylinders, and both sets of points work together. With this arrangement, the fouling of a plug, or failure of one coil or set of points will not result in the "loss" of one cylinder. The other system will still be working away, and while the engine does not run as smoothly on one set of plugs as it does with two, it will run and produce plenty of power. Also, by having the coil pushing high-tension current across one plug in a high-pressure cylinder, and then having the other under low pressure (the non-firing cylinder would be between exhaust and intake strokes) less voltage is required to fire both plugs. Incidentally, the coils are rubber-mounted, as are many other miscellaneous bits of hardware on the bike.

With this unusual ignition system, it has been found that maximum power, coming in at 7500 rpm, is delivered with the ignition set at 30-degrees of advance. But, the best power curve, slightly less at the

been running the same followers for many, many racing laps, there is no sign of scuffing or wear.

Most of the valve gear is stock. The rockers have been lightened slightly (nothing drastic, it should be understood) and polished, but not to a mirror finish. Webco competition valve springs are used, and light-alloy retainers, with Porsche cushion-type washers under the springs.

A pair of perfectly standard 1-3/16" Amal Monobloc carburetors are fixed to the head's inlet stubs by a vibration-insulating rubber-hose connection. Long intake trumpets have been fitted on these to smooth airflow into the carburetors and to add length to the intake tract. During the Ascot TT, 460 main jets were used, and it is a tribute to the efficiency of this engine that after 100 racing laps, 1-1/2 gallons of gasoline remained in the 4-gallon tank. That is a little over 23 mpg under full-throttle racing conditions, with the engine screaming along at somewhere near 7000 rpm most of the time.

No real attempt has been made to "tune" the intake and exhaust system. The

In all, the Harman and Collins/Triumph TT bike is a remarkable piece of machinery. All too often, modifications of such a radical nature do not prove out very well, but then most radical modifications are made by people who have only a hazy notion of what they are trying to accomplish. When this sort of thing does work, it is because the "tuner" has a good working knowledge of fundamentals and does not simply hack away for the sake of change. Gary Robison, cam grinder *extraordinaire* does know what he's doing, and he has done it very well. •

BY GORDON H. JENNINGS

Triumph "TT" Special (650 cc OHV twin)
* Limited-production competition model

BLUEPRINT OF A CHAMPION

Take a look at the record of records racked up by Triumph's "TT" Special.

★ WINNER 50-LAP TT NATIONAL CHAMPIONSHIP, GARDENA, CALIFORNIA 1964
★ WINNER CALIFORNIA STATE CHAMPIONSHIP TT, GARDENA, CALIFORNIA 1964
★ WINNER NATIONAL CHAMPIONSHIP HILL CLIMB, MUSKEGON, MICHIGAN 1964
★ WINNER NATIONAL CHAMPIONSHIP SCRAMBLES, BUTLER, PENNSYLVANIA 1964
★ WINNER 500-MILE NATIONAL CHAMPIONSHIP ENDURO, RED MOUNTAIN, CALIFORNIA 1964

Following records set at Bonneville Salt Flats, Utah

★ SPEED RECORD HOLDER A.M.A. CLASS SC 205.785 MPH, 8/21/62
★ SPEED RECORD HOLDER A.M.A. CLASS AA 159.54 MPH, 8/25/61
★ SPEED RECORD HOLDER A.M.A. CLASS PSC 137.075 MPH, 8/21/63

In case you need further proof, the "TT" Special is powered by the same engine that established Triumph's Bonneville as the world's fastest motorcycle—230.269 mph!

ALL RECORDS CONFIRMED BY AMERICAN MOTORCYCLE ASSOCIATION, COLUMBUS, OHIO.

*This one's hard to get, but worth waiting for. The best ones always are.

Western Distributor/JOHNSON MOTORS, INC. 267 W. Colorado, Pasadena, California 91102

TRIUMPH

Eastern Distributor/TRIUMPH CORPORATION, Towson, Baltimore, Maryland 21204

CW Trail Test TRIUMPH MOUNTAIN CUB

OUR TRAIL TEST machine this month pretty well fills the bill of what we have preached for sometime; a real motorcycle, slightly undersized, properly equipped and geared, is the best trail machine when all demands are given consideration. For those who desire only a trail scooter type bike, or for those who want a motorcycle that is well suited to the problems of off-the-road work, Triumph's Mountain Cub stands among a very small and select group of machines.

Several trail scooters on the market may possess virtues not to be found in larger motorcycles such as lightness and extreme portability, and though the Cub's 250 pounds doesn't exactly qualify it as a lightweight, it can be easily handled by the average rider in all but the steepest or roughest territory where picking the machine up bodily and carrying it over an obstacle is the only answer. A 7½ inch ground clearance helps a little at those times, too.

Triumph thoughtfully installed their "trials" gearbox in the Mountain Cub, offering two very low and close together first and second gears, with higher gear ratios in third and fourth with a wide spread between second and third. This simply means that the first two gears are low for slow and difficult going, and third and fourth are better suited for cruising speeds on dirt roads and across smoother terrain. It works well until one tries to race in a scrambles or some such event and finds that he is either revving the very devil out of it, or lugging it with equally disastrous consequences.

Obviously the rider who wants to race the Cub should buy the Sports Cub scrambler, a better machine for the job for several reasons, not the least of which are the gearbox, increased horsepower, and its lighter weight due to the absence of the lighting equipment used on the Mountain Cub. Since this evaluation is being written to fit the viewpoint of the trail rider, not the scrambles racer, or even the part-time racer, we can easily overlook a point or two we might come to odds with.

Rated at 15.1 horsepower @ 6,500 rpm, the single-cylinder, overhead valve, four-stroke engine, is built in unit with the four-speed gearbox, all in alloy and handsomely polished to a tasty finish. The exhaust pipe is carried close to the engine and high, passing under a special bend in the tubular frame which further tucks it in out of the way. One fall to the right proves the value of this kind of planning. A 300 x 19 Dunlop Trails Universal tire is mounted on the front, with a 3.50 x 18 rear. Dunlop Sports "knobbies" can of course be fitted; some might prefer them under certain conditions.

Suspension is by hydraulically dampened telescopic forks and shock absorbers. Steering is very quick, an asset when stumbling around in rugged country trying to avoid rocks, fallen trees, holes, etc., and the ride is downright comfortable. An interesting final thought is the inclusion of a paper element air cleaner, a handy device.

Nearly every feature anyone could want, and one he does not want or need, is either standard, or available as an accessory, such as the chrome plated carrying rack made by Accurate Accessories. The price seems high only if you have been shopping the trail scooters or lighter and smaller machines; the advantages are worth the difference. ●

CYCLE WORLD ROAD TEST

TRIUMPH THUNDERBIRD

LONG BEFORE a certain automobile manufacturer in Dearborn cribbed the name, "Thunderbird" meant total performance to thousands of motorcycle enthusiasts. To them, Thunderbird meant only one thing: Triumph's very smooth and rapid 40 cubic inch vertical-twin motorcycle, which was quite justly regarded as being *the* machine for the sporting rider. Times have changed (the "40-inch 'Bird" was introduced 14 years ago) and there are now faster Triumphs, but the Thunderbird is still an important part of the Triumph line, and it is still a remarkable package of performance and reliability. As a matter of fact, the many improvements introduced over the years have made it even better than before — and that is saying a great deal.

In the 1964 version, the Thunderbird incorporates all of the improvements bestowed on the other big Triumph twins, such as the unit-constructed engine and single down-tube frame of recent years. Most people are at least somewhat aware of the overall changes in the Triumph, but there have been detail changes for 1964 that may have been overlooked. For instance, there is the redesigned oil scavenging system, which has a larger sump area to improve scavenging and thus reduce oil drag on the flywheel at high engine speeds. In the same general area, the breather has been reworked to improve crankcase ventilating, and the hose from the breather has been re-routed up to the oil supply tank, where it leads into the de-frothing tower. With this arrangement, any oil that might find its way out of the breather is captured in the supply tank, instead of being lost. Also in the interest of oil conservation, the Thunderbird has new slotted oil control rings, which should be more effective in scraping from the bores oil thrown up from the connecting rod journals. The pistons themselves are being fitted with a bit less clearance than before to give more quiet running. These pistons are, incidentally, slotted so they will not

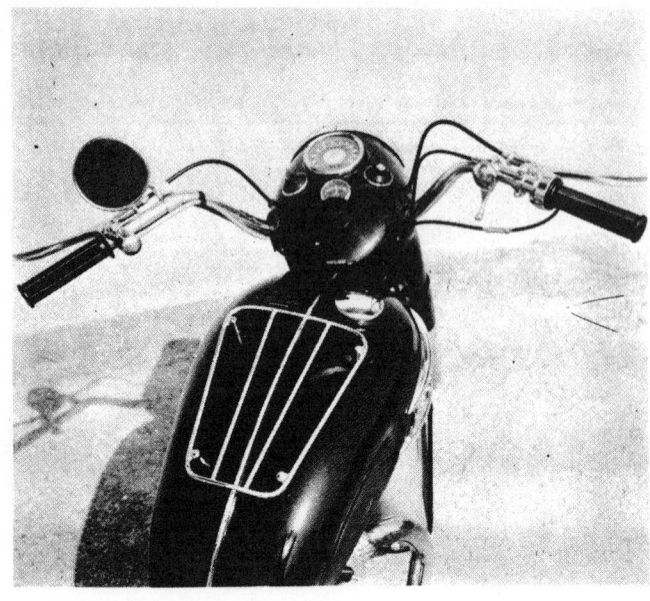

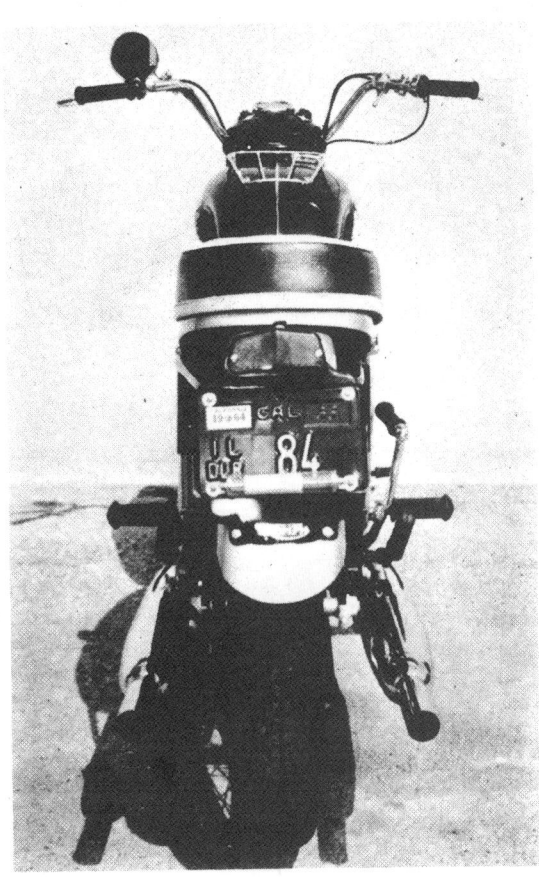

"outgrow" the bore if the engine is overheated.

One very important feature of the Thunderbird, not shared by other Triumphs, is the new 12-volt electrical system. Current is supplied by an alternator, as on the other Triumphs, but after passing across a silicon crystal full-wave rectifier, it is fed into a pair of 6-volt batteries, connected in series to give 12-volts for the ignition and lighting systems. Voltage regulation is handled by a zener diode transistor mounted in an aluminum plate (for heat dissipation). This 12-volt system does not really do anything appreciably better than a 6-volt system, but it will do it a lot longer, and that certainly justifies the change.

Another change in the Thunderbird (and the other Triumph twins) for 1964 is a completely new set of forks. Previous Triumph forks had long, small-diameter coil springs located inside the fork tubes, and damping was handled by oil squishing past a metering rod. The new forks have much shorter springs, with larger diameter coils, located outside the upper fork legs above the "sliders." In this position, they do not "scuff" along the fork leg as much as the earlier internal springs, and the ride is thus improved. Also, the new forks have considerably better damping than earlier Triumphs, and not only has this improved the ride, but it has done wonders for the handling as well. The machine feels so secure when cornering hard that one leans deeper and deeper into the turn — and then curses the stand for hanging down where it can drag.

Those who prefer pounding out through the rough will be pleased to know that there is a new, and more effective, oil seal on the forks. This seal is of the double-lip type, with a spring-backed lower lip to keep the oil in, and an upper lip that serves as a scraper to keep dust and moisture away from the lower lip — which increases its effective life. The people at Triumph swear that this new seal is absolutely effective, and that no oil loss is possible under any conditions. Time will tell; we certainly never noticed any oil escaping during the time we had the Thunderbird, and we rode it well over 1000 miles under all conditions. The brush-banger type rider will also appreciate the new travel-stop at the bottom of the fork. Each leg has inside its lower end a long, tapered peg, and if the forks are driven up to the limit of their travel, this peg moves into a hole in the upper stanchion. As the peg moves into this hole, it gradually closes off the upward fork travel that makes it virtually impossible to actually "bottom" the forks.

No mention is made by Triumph of any alterations to the bike's brakes, but we are sure that there has been at least a change in lining material. The brakes on the Thunderbird seemed to be more powerful than those on previous Triumphs we have tested, and they were definitely smoother in their operation. There was none of the faint, distressed shuddering we have noted in stopping from high speeds that we have experienced with earlier "big" Triumph twins. It is possible that this is due to the peculiarities of individual machines, but for whatever reason, the brakes on the new Thunderbird were exceedingly good.

Lucas's new magnetic-drive speedometer is used on the Thunderbird, and it has an especially steady needle, which makes it easy to read. A trip-meter (which you re-set by twirling a small knob) is provided in the speedometer head, in addition to the usual odometer. Unfortunately, the fact that the speedometer gives steady readings does not also mean it is accurate. There is an 11.3 percent error (slow) at 30 mph, and although the error drops as speed increases, an indicated 70 mph is still only 68.7 mph. Of course, most motorcycle speedometers err even more, and what makes it even worse is that the error usually gets larger as the speed goes up. By that standard, the Triumph's speedometer is a marvel of accuracy.

Accessibility for routine service is quite good on the

Sans bodywork the T-Bird loses none of its attractiveness.

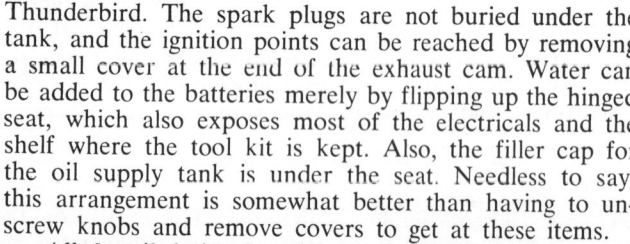

Thunderbird. The spark plugs are not buried under the tank, and the ignition points can be reached by removing a small cover at the end of the exhaust cam. Water can be added to the batteries merely by flipping up the hinged seat, which also exposes most of the electricals and the shelf where the tool kit is kept. Also, the filler cap for the oil supply tank is under the seat. Needless to say, this arrangement is somewhat better than having to unscrew knobs and remove covers to get at these items.

All the oil drain plugs for crankcase, transmission and primary drive case are easily reached, and those who have struggled with draining the oil supply tank, which required the removal of a plug to which the oil outlet was fitted, will be delighted to hear that there is now a separate oil drain plug — easily reached on all the Triumphs but the Thunderbird. The enclosure in front of the rear wheel, which keeps the engine and rider so clean and neat, completely covers the oil tank, and it must be unbolted (3 bolts, 2 screws and a knob) before the drain plug is within reach. We rather imagine that many Thunderbird buyers will elect to simply leave the enclosure in the garage; the full fender is there anyway. This would leave the air cleaner completely exposed (and it is a very good, fiber-type air cleaner, by the way) but that will not reduce its effectiveness.

The best of the Thunderbird's many qualities (performance, economy, reliability, etc.) is that it is such a civilized motorcycle. The mildly tuned engine is phenomenally smooth, because of the single carburetor and low, 7.5:1 compression ratio, and it will idle and pull at low speeds in a way that makes it a particular pleasure to ride in city traffic. And, of course, it is also capable of boring down the open highway at a most impressive rate.

We simply cannot find fault with the ride, or riding position, or controls. Everything operates with great ease and precision, and the seat, pegs and handlebars suited us perfectly. Actually, we suppose the bars are a bit too high for high-speed cruising, and a bit low for riding in town, but they constitute a compromise that is not likely to be improved upon so long as the Thunderbird is used in all kinds of riding.

It is impossible to say too much good about the Thunderbird's starting characteristics. Hot or cold, the engine would almost invariably come to life with a single kick — and not a particularly vigorous kick at that. No special starting "combination" is required, either; the engine has an automatic advance/retard on the ignition (as do many others) and so when it is warm, the rider need only tweak the throttle a hair and crank it through. When cold, the drill is the same except that the choke is used. There is no need to "tickle" the carburetor float.

In all, the Thunderbird impressed us as being very much the gentleman's motorcycle. It starts too easily to force one to work up a sweat, and it proceeds to its destination smoothly and quietly. There is ample performance on tap, and while it is not much faster than some of the hotter small-displacement bikes, it does what it does without a lot of gear-stirring and threshing of the engine. We return the Triumph Thunderbird with reluctance, for it has been as pleasant a motorcycle as ever offered to us for test. May Triumph sell a million of them. •

TRIUMPH THUNDERBIRD

SPECIFICATIONS

List Price	$1,066 POE
Frame Type	tubular, single loop
Suspension, front	telescopic fork
Suspension, rear	swing arm
Tire size, front	3.25-19
Tire size, rear	4.00-18
Break lining area, sq. in.	32.5
Engine type	vert. twin, ohv
Bore & stroke	2.79 x 3.23
Displacement, cu. in.	39.6
Displacement, cu. cent.	649
Compression ratio	7.5:1
Bhp @ rpm	34 @ 6300
Carburetion	1 1/8" Amal Monobloc
Ignition	battery and coil
Fuel capacity, gal.	5.0
Oil Capacity, pts.	6.0
Oil System	dry sump
Starting system	kick, folding crank

POWER TRANSMISSION

Clutch Type	multi-disc, wet plate
Primary drive	duplex chain
Final drive	single-row chain
Gear ratio, overall:1	
4th	4.84
3rd	5.76
2nd	8.17
1st	11.8

DIMENSIONS, IN.

Wheelbase	55.5
Saddle height	31.5
Saddle width	10.2
Foot-peg height	13.0
Ground clearance	7.7
Curb weight, lbs.	400

PERFORMANCE

Maximum practical speed (after 1/2-mile run)	91
Max. speed in gears @ 6500 rpm	
4th	103
3rd	86
2nd	60
1st	40
Mph per 1000 rmp, top gear	15.8

SPEEDOMETER ERROR

30 mph, actual	26.6
50	48.0
70	68.7

ACCELERATION

0-30 mph, sec.	2.0
0-40	2.7
0-50	4.7
0-60	6.6
0-70	10.0
0-80	13.6
0-90	22.2
0-100	
Standing 1/4 mile	15.4
speed reached	84

ENGINE / ROAD SPEED

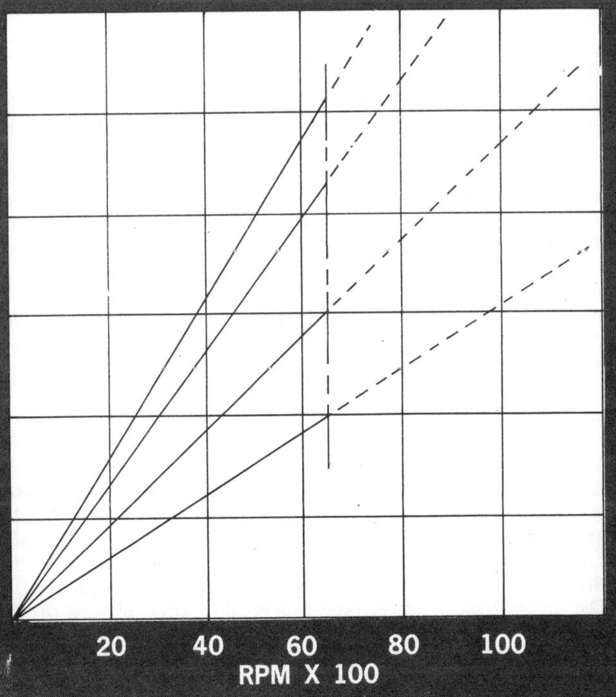

ACCELERATION

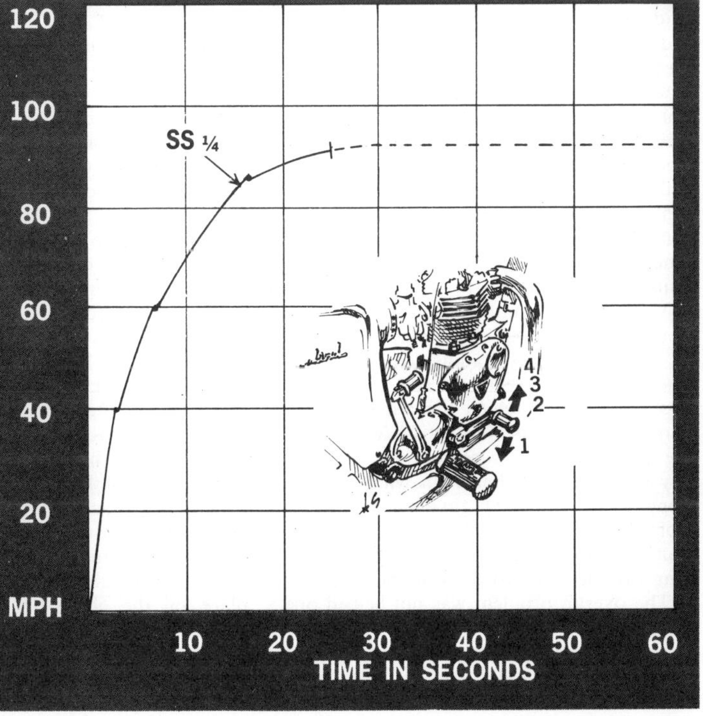

WE ~~TRIUMPH~~ try AT BONNE-VILLE

or: Just wait til last year

Record holders a pair, the Richards/Cycle World 650 Class A, and Gordon Jennings' 650 Class C.

Jennings on the CW bike, and Gary Richards on Richards' record holder.

As many of you will recall, the CYCLE WORLD staff went to the Bonneville Speed Trials in 1963 and captured the Partial Streamlined, Class-C, 650cc record with a very slightly modified Triumph TT Special — raising the record by 10 mph in the process. Our two-way average of 137 mph was by no means the fastest any Class-C motorcycle has ever run up at the salt flats, but it was enough to get the record and enough to be a lot of fun. Accordingly, we decided that a return to Bonneville in 1964 was in order, because it seemed obvious that we could get up to a much more impressive speed with a more elaborately prepared machine. Also, there was the thought that as we had taken the record so easily, someone else would surely grab it away if we were not there to defend it. We were wrong on all counts: none of the motorcycles that ran for our record in the 1964 Speed Trials came even close to our 1963 speed — and, indeed, we were unable to match last year's speed ourselves. In short, friends and readers, we trotted up to the salt in 1964 and laid a big, fat egg. No record; no speed; no nothing.

Of course, there is a bit more to the story than that. Actually, we were humbled by circumstances, which prevented us from making more than 4 runs during the entire speed trials, and which also prevented us from doing any testing before going out to the salt flats. The lack of tune-up runs can be attributed to the condition of the salt and to weather: the salt was so soft and rutted that the organizers were forced to close the course periodically and send out grading equipment to smooth the bad sections, and there were two days during which the wind blew so hard that no runs were possible. In fact, the last runs made were very early on Friday morning, and then high winds (which stopped all running) finally blew water from a nearby "lake" over the course and shut everything down — effectively chopping off the last two days of the meet.

The overall problem was not helped one bit by our own tardiness. As is customary in racing efforts, nothing was even started until the last minute, and then delays in obtaining a set of experimental pistons used up what little time remained. In the end, the final touches were put on the motorcycle at 8:00 A.M. on the Sunday that marked the opening of the 1964 speed trials — and we still had to transport the bike some 750 miles. Consequently, the first day of the running was lost to us entirely. The rest of the meet was marked by frustrating delays while we waited for course conditions to improve, and twice we reached the line, ready to make a run, only to have something go amiss which closed the course for the rest of the day. We should make quite clear that this was not the fault of the SCTA, who are the organizing body at the speed trials; weather and the general condition of the salt are not things over which they have any control.

All of this was doubly frustrating because, this year, we had prepared a machine that had all the potential to push our record up to around 150 mph. Last year, after we had set the record, a fire destroyed our TT Special, and it had to be entirely rebuilt for 1964. While we were at the task, a lot of changes were made. Among these was to modify the original Triumph front hub so it could be used at the rear. The wheel bearings are identical, so the standard rear axle would fit nicely in the front hub, and spacers were machined to locate the hub in a centered position inside the rear swing arm. A spacer was also fitted inside the hub, between the two bearings, to give the bearings the proper pre-load when the axle nuts were pulled down tight. Obviously, if a front hub is to be used in a rear wheel, some provision has to be made for mounting a sprocket, and this was done by machining 6 holes (on a 4-inch circle) through the two conical plates that form the center of the hub. Into these holes, 6 short pieces of heavy-wall tubing were inserted and brazed (using a special high-strength chromium-bronze rod) into place. These tubes extend out from the hub far enough to hold the sprocket in place on the chain line. The stock front brake mechanism was left intact, and a piece of rectangular bar stock was welded on the inside of the swing arm to engage a slot in the brake backing plate — which handled braking torque. The brake is cable operated from a handlebar lever. This special hub was necessary because the stock rear hub has the drive sprocket around the brake drum, and the sprocket cannot be made small enough to get the gearing required for very high speeds.

Because a front brake is not needed, or even desirable, at Bonneville, we again borrowed one of Rich Richards' brakeless front wheels and high-speed Avon tires. We could have used one of these special tires at the rear, too, but Avon is no longer making them and as a result we fitted one of the now-obsolete low-hysteresis Avon road racing tires on the rear wheel. The new high-hysteresis road racing tires made by Dunlop are great for cornering on pavement, but the same high internal friction that makes them cling to the road surface so ferociously also makes them absorb a lot of power and overheat when run at very high speeds. The new tires are magnificent in the sort of application for which they are intended, but we had something very much different in mind.

The fairing, seat and fuel tank we used last year were destroyed in the fire, so these items had to be replaced. A smaller fuel tank was obtained, and a smaller, lower seat as well. Because of difficulties we had experienced last year in having to remove the fairing for the otherwise minor task of adjusting the ignition advance (and the more major operations of changing countershaft sprockets and stripping the cylinder head) we decided to go to a 3-piece fairing with removable side panels. Custom Plastics, who supplied the tank and seat, also provided the fairing, and while we blessed them for the first pair of items, the fairing was far from right. It simply didn't fit the bike, and this small problem was further complicated by the fact that the fairing had apparently warped as the resin cured, leaving one side somewhat higher than the other. We eventually pulled it into shape with the brackets, but time was lost that could have been more profitably spent on other phases of the project. Also, the frontal area of the new fairing was too large; we would have done better with the narrow, one-piece fairing that we used before — which was,

Honda megaphones, Custom Plastics seat, tank and fairing, give CW Bonneville a distinctive look.

38mm Dellorto carburetor and plastic timing case breather hose can be seen on right of CW special.

incidentally, also made by Custom Plastics and which was entirely satisfactory.

Although we did change this and that around the motorcycle, it was in the engine that most of the progress was made. The Triumph engine is, of course, basically sound, and no really drastic changes are required, but there are many refinements that can be made to boost power output. First, we replaced the stock 1 3/16" Amal Monobloc carburetors with a pair of TT-pattern Dellortos; these having a throat diameter of 35mm, or 1.378". The Dellorto carburetors will not simply "bolt-on" and we had no desire to do this in any case. Instead, the standard thread-in

"Rich" Richards, master of the Cycle World pits and builder/tuner of the 650 Triumph fuel burner.

carburetor stubs were removed and longer aluminum mounting stubs welded into the head. The ports had to be machined considerably oversize for a short distance inside the head to accommodate the special stubs. The special stubs were bored to a double taper inside, starting with a 35mm diameter at their outer end, narrowing down to 1 3/16" at the point where they enter the head, and then opening up to almost 1 1/2" where they reach the back of the valve seat.

This arrangement creates a venturi in the port just upstream from the intake valve (which was 1 5/8" in diameter) that increases gas velocity and provides an inertia ram effect in the medium-high speed range. Maximum power would have been obtained with straight-through porting, but as the rules restrict the larger Class-C bikes to 4-speed transmissions, we needed a reasonably wide spread of power for Bonneville. One must have exceedingly "tall" gearing to get the required speed, and if the engine's power range is too narrow, it will simply not be able to pull the motorcycle fast enough to climb up "on the cam" in top gear. From the standpoint of theory, and from practical experience, we knew this to be true — and the entire engine was set up accordingly.

On the exhaust side of the cylinder head, we used stock valves and left the ports virtually untouched. However, instead of last year's stock 1 3/8" diameter upswept straight pipes we employed a set of Webco's 1 3/4" TT-style exhaust pipes and capped them with Honda megaphones. The megaphones are made by Honda for road racing versions of their Hawk, and we simply carved a larger inlet in these and stuffed them onto the 1 3/4" Webco pipes. We had planned to make a set of megaphones, but the Honda parts were so near what we wanted that there seemed little point in manufacturing a set from scratch. The overall length of the exhaust system was, of course, "tuned," as was the intake system.

One of the major internal changes in our Bonneville engine for 1964 was a switch to Harman & Collins roller tappet cams. The roller tappets are carried in special aluminum alloy tappet blocks, and as these are slightly larger in diameter than the stock parts, the cylinder block holes in which they fit had to be enlarged and the lower three cooling fins on the block had to be trimmed back. The valve timing with the roller tappet setup has the intake valve opening 50-degrees before top center and closing 80-degrees after bottom center. The exhaust valves open 79-degrees before bottom center and close 45-degrees after top center. Lift for the intake valves is .422"; the exhaust valves lift .378". The valves are "off the shelf" Triumph parts, but the valve closure is done with special S&W racing valve springs and light-alloy retainers. The pushrods are of aluminum, like the stock parts, but are hollow and have only one steel end; the upper end is formed into a cup and rides directly against the ball at the tip of the rocker.

As we have mentioned before, detonation was a major problem last year, and while casting about for some means of reducing the tendency to detonate we discovered Webco's "Power Sphere" piston. After our earlier experiences with the Triumph, we had decided that what was really required was a piston with a relatively low crown that extended out to the sides to form squish areas adjacent to the valves. By doing this, the end-gasses trapped in the area opposite the spark plug are squeezed down to a thin layer that loses heat to the head too quickly to auto-ignite. Unfortunately, Webco had the Power Sphere piston (which gives the squish area just described) available only for the 500cc Triumph; pistons for the 650cc Triumph were nearing production status, but at the time we inquired the tooling had not been completed. Still, we wanted these pistons badly, and to accommodate us Webco's suppliers produced a special set in advance. This was, of course, a great deal of bother for them, and even proceeding with all due haste they could not promise delivery more than a few days before we were scheduled to leave for Bonneville. Even so, these promised to be the pistons that would do what we wanted and we therefore elected to wait.

As it developed, the wait was worth all the ulcers and gnawed fingernails it produced. With the ghastly sound of last year's detonation still ringing in our ears (or so it seemed when we were making plans) it was decided that about 11:1 compression was all that the engine would stand. However, the makers of the pistons had great faith and in due course (a bare 72 hours before D-Day) delivered a set of 12:1 pistons. These were installed with some misgivings on our part, but with the time dwindling away we had little choice. Frankly, we expected that we would again encounter problems with detonation — less severe, of course —

CONTINUED ON PAGE 80

ONCE YOU PENETRATE the dazzle of chrome plating, the most notable thing about Bob Belanger's custom '50 Triumph is the far out front forks. A 7" addition to the wheelbase effectively lowers the frame, which was cut to angle the steering crown. Some loss of low speed handling resulted, but stability at speed is reportedly enhanced. This is bike number three for Bob, co-owner of Belanger Bros. Muffler Shop in the Los Angeles suburb of West Covina. Starting from scratch, the customizing took four months and 1400 dollars. The 40" engine is basically stock, but like the frame itself, is heavily plated and polished. The aluminized cylinders transfer heat faster, look different than the stock cast iron. Engine accessories include rocker cover caps and cast, finned oil line between rocker boxes. Monobloc carb updates twin to current specs. Baldwin Park Bike Shop assisted on the project and fine lacquer flames are by Reyes. •

"Putting its best wheel forward"

PHOTOS BY CAL WEST

HIGH ADVENTURE LINE FOR '65

T120/TT   BONNEVILLE SPECIAL COMPETITION
Twin Carburetors 40 cu. in (650 c.c.)

For the competition expert who wants top performance with Twin Carburetors without lighting equipment, primarily for off the road, racing and competition use.   Color: New Pacific Blue and Silver.

T120/R   BONNEVILLE ROAD SPORTS
40 cu. in. (650 c.c.)

The fastest standard motorcycle in the world today, with unitized engine-gearbox construction. For the expert rider.   Color: New Pacific Blue and Silver.

TR6S/R   TROPHY ROAD SPORTS
40 cu. in. (650 c.c.) Single Carburetor

Smooth and reliable, a truly high performance winner.   Color: New Burnished Gold and Alaskan White.

T100S/C   TRIUMPH TIGER 100 COMPETITION TROPHY
30.5 cu. in. (500 c.c.)

The top choice for Woods, Enduro and Club competition. The Famous "Jack Pine" Model.   Color: New Burnished Gold and Alaskan White.

T100S/R   TIGER ROAD SPORTS
30.5 cu. in. (500 c.c.)

More than a tiger in your gas tank. It's all tiger. All latest improvements for '65.   Color: Burnished Gold and Alaskan White.

T20   TIGER CUB ROAD MODEL
200 c.c. Lightweight

Full powered 4 cycle OHV lightweight performance in the Triumph tradition. Reliable, easy to start, economical.   Color: Flamboyant Scarlet and Silver.

T20S/R   TIGER CUB ROAD SPORTS
200 c.c. Lightweight

High performance, OHV 4-cycle single cylinder engine. Has new competition type front forks.   Color: New Pacific Blue and Silver.

T20S/C   TIGER CUB COMPETITION SPORTS
200 c.c. Lightweight

The competition model of the famous Tiger Cub, OHV 4-cycle with many new features for 1965. Polished Aluminum fenders.   Color: New Hunting Yellow.

OTHER MODELS AVAILABLE   6T THUNDERBIRD 40 cu. in. (650 c.c.) □ TR6S/C COMPETITION SPORTS 40 cu. in. (650 c.c.)
T20/J JUNIOR CUB 5 bhp. □ T120/C BONNEVILLE COMPETITION SPORTS

Note: Western specifications may vary from those shown.

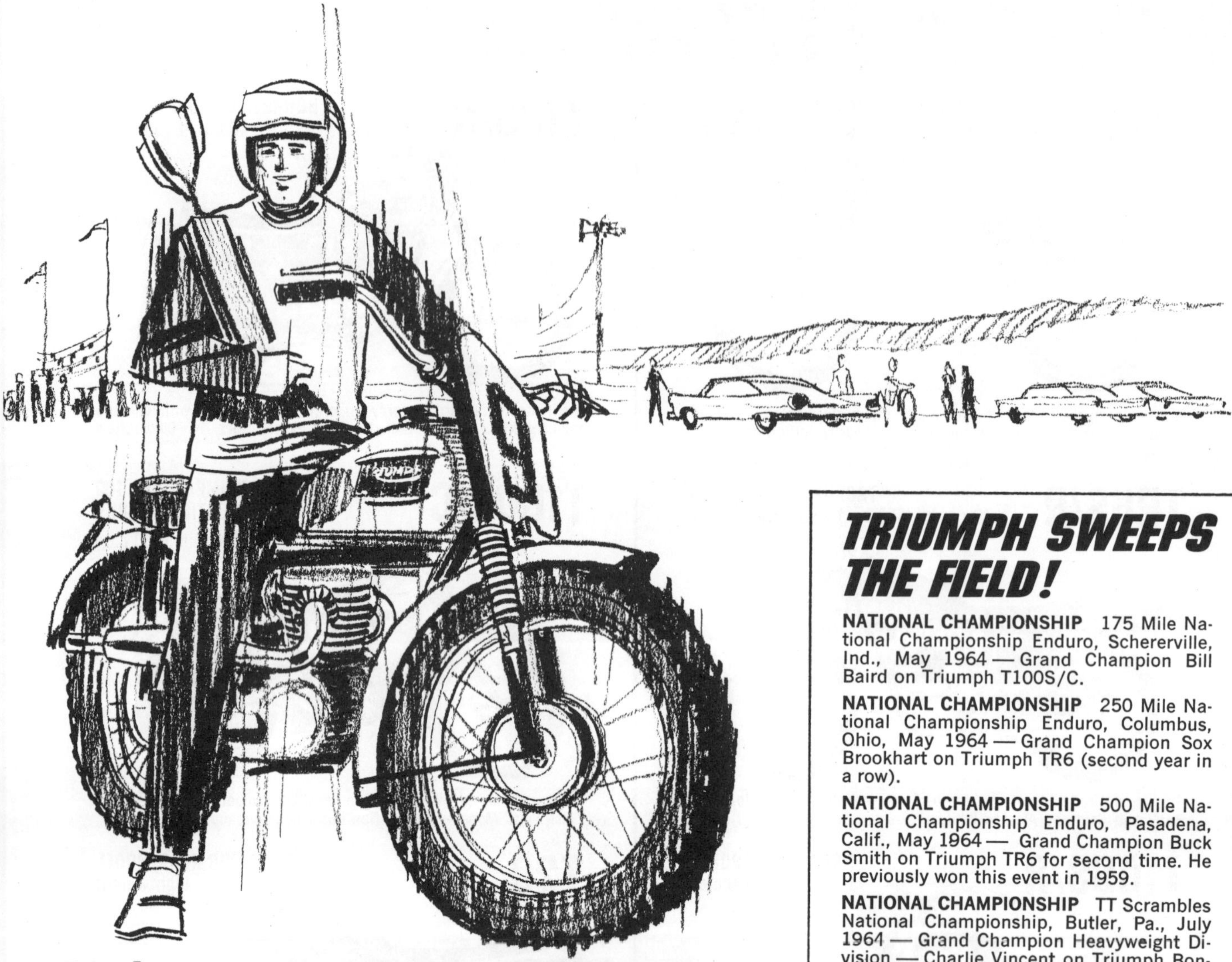

DON'T HANDICAP YOURSELF!

Every rider wants to get the most fun out of his motorcycling dollars. That's why you should look carefully before you buy. Look past advertising claims and you'll find a Triumph is your best buy — dollar for dollar. The Triumph power and reliability spell **fun** for you wherever you go, be it on the open road or in competition riding.

Triumph operates a parts service, second to none, coast-to-coast. Speed tuning, competition and racing parts are available for Triumph twins and lightweights. Also accessories are available from your Triumph dealer.

Triumph continues to outsell all other vertical twins in the U.S.A.

Your Triumph dealer invites you to inspect and test-ride the new Triumph line for 1965.

For FREE Catalog in full color and name of dealer nearest you, Write Dept. K

Johnson Motors, Inc.
WEST . . .
P. O. Box 457, Pasadena, Calif. 91102

The TRIUMPH Corporation,
EAST . . .
Towson, Baltimore, Maryland 21204

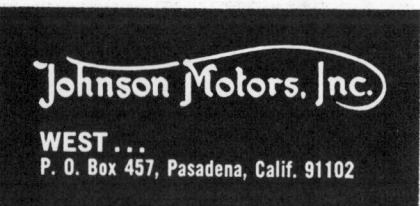

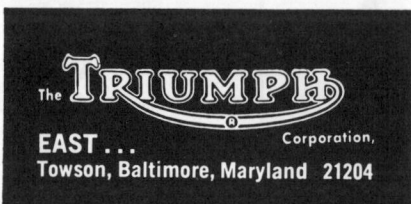

TRIUMPH SWEEPS THE FIELD!

NATIONAL CHAMPIONSHIP 175 Mile National Championship Enduro, Schererville, Ind., May 1964 — Grand Champion Bill Baird on Triumph T100S/C.

NATIONAL CHAMPIONSHIP 250 Mile National Championship Enduro, Columbus, Ohio, May 1964 — Grand Champion Sox Brookhart on Triumph TR6 (second year in a row).

NATIONAL CHAMPIONSHIP 500 Mile National Championship Enduro, Pasadena, Calif., May 1964 — Grand Champion Buck Smith on Triumph TR6 for second time. He previously won this event in 1959.

NATIONAL CHAMPIONSHIP TT Scrambles National Championship, Butler, Pa., July 1964 — Grand Champion Heavyweight Division — Charlie Vincent on Triumph Bonneville.

NATIONAL CHAMPIONSHIP Hillclimb, Mt. Garfield, Muskegon, Mich. — August 1964 — Grand Champion Joe Hemmis on Triumph Twin (650 cc) for third time. Joe previously captured this title in 1959 and 1962.

NATIONAL CHAMPIONSHIP 150 Mile National Championship Enduro, Cayuta, N. Y., Sept. 1964. Also two-time Grand National Champion U. S. Enduro Rider for 1962 & 1963 — Bill Baird on Triumph T100S/C.

NATIONAL CHAMPIONSHIP 20 Mile National Championship Dirt Track Race, Sacramento, Calif., Sept. 1964 — Gary Nixon on Triumph T100 Dirt Track Racer. Also 2nd in 200 Mile National Championship Road Race, Daytona, Fla., March 1964.

JACK PINE ENDURO Traditional Jack Pine Enduro, Lansing, Mich., Sept. 1964 — Grand Champion Roger Kussmaul on Triumph Cub. Score: 994.

INTERNATIONAL SIX DAY East German, September 1964, I.S.D.T. **(International Six Days** Trials) — U. S. Riders Dave Ekins — T100 Gold Medal; Cliff Coleman, TR6, Gold Medal; John Steen, T100 Silver Medal.

NATIONAL CHAMPIONSHIP Gardena, California, September 26, 1964 — Triumph TT Specials led by Dave Palmer sweep first 7 places in 50 lap H/W National Championship TT Race.

a

THREE-WAY TRIUMPH
Drags, Custom Shows and Road Race wins — all in a year's work.

b

Many special parts were made by the owner, including centermount oil tank incorporating paper air filter elements and rebaffled Matchless mufflers. Filled frame scores extra points in custom shows. **A**

The triple-threat custom is finished in sungleam red and Aleutian gray lacquer, took five months to build. Double leading shoe front brake was added for road racing, a skillful improvement over standard brake. The already powerful 650cc TT Special engine was overbored .040 and fitted with 9.5:1 forged pistons. Q cams operate big valves through lightweight valve train. Lightened flywheels reduce 1/8-mile time. **B**

Perched in contrast upon the gleaming rump of Harry Penn's 1963 TT Special, its battered Pennsylvania license plate tells part of this motorcycle's story. **C**

BY CHUCK CLAYTON

c

MOST OF THE MOTORCYCLES we get to try here at CYCLE WORLD are shiny new and full of promise, but occasionally we come across an old acquaintance like Harry Penn's 1963 Triumph TT Special, which has led a busy life this past year, yet feels almost as good, and in some ways better, than when it came out of the box.

Penn has trophied a dozen times on the bike at Eastern 1/8-mile drag meets, has won top awards in custom shows with it, and even herded it to second place in an AAMRR road race. He then rode the remarkable road eater from his home in Philadelphia, Pa., to Southern California, where all good Triumphs get to go, and where Penn now works, for a Triumph dealer naturally.

Penn generously offered us a ride on his pride, which started easily, partly thanks to the 9½-to-one pistons which Penn runs in preference to the stock 12-to-ones. The bike accelerated strongly enough to satisfy the most impatient street rider but flattened rather early, with the symptoms of spark failure. There was very little vibration up to that point, however, and the cycle still cruised comfortably at freeway speeds in top gear. For the total of $1900 which Penn spent for the machine, including purchase price, he has made an investment that is still paying dividends. ●

TRIUMPH TR-6 SR

CYCLE WORLD ROAD TEST

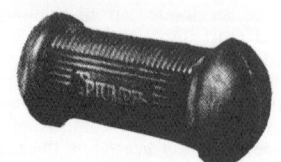

WE HAVE A GAME around CYCLE WORLD's offices, and it is called "capsule commentary," in which all of us try to reduce whatever a motorcycle is to a short (and hopefully, succinct) phrase. The Triumph TR-6 SR "Highway Trophy" came out as The Road-Riders' Delight, and that just about sums it up nicely. It is not the fastest motorcycle in the world, or even the fastest Triumph, for that matter, but it is definitely one of the best touring machines ever delivered into our irreverent hands.

We are told that many people who buy Triumphs (and this must take in a fairly substantial percentage of the motorcycle buying public) choose the T-120R, which is the twin-carburetor "hot" version of Triumph's 40-inch bike. That is understandable, as that particular motorcycle is a prestige item, with bags of horsepower. Even so, these buyers who are basically touring riders are making a mistake with the Bonneville T-120R: the TR-6 SR is easier to start, smoother-running, and has the sort of low-speed torque that makes it a pleasure to ride in areas where traffic density makes pure speed quite useless.

Actually, the differences between the hot and merely warm versions of the Triumph engines is quite small. The hot model has twin carburetors and relatively zingy valve timing, but apart from that the engines are the same. The TR-6 SR has a single carburetor, but this is not so great a handicap, in terms of power, as one might think. Having its crankpins together, and equally spaced power pulses, the Triumph has its intake strokes spaced 360-degrees apart, and this means that the carburetor only feeds one cylinder at a time — while the other awaits its turn. So, insofar as the cylinders are concerned, there are two 1 1/8-inch carburetors; the only thing that is lost is the "supercharging" effect of sound waves in the intake system, and this is important primarily at high engine speeds. In low and medium-speed running, the single carburetor (which is operating under the advantage of more or less constant-flow conditions) gives better throttle response, better fuel economy and, as we have said, easier starting. Also, and not the least of the advantages by any means, the single carburetor will stay in "tune" almost indefinitely, as there will be no problems with mismatched adjustment of throttle slides or float levels.

While on the subject of easy starting, it is worth mentioning that all of the Triumphs we have had for test purposes came to life with remarkably little urging; but the TR-6 SR established some sort of a record. Seldom were more than two kicks needed to get it running, even when cold, and more often than not a single whack at the kick-lever did the trick. In fact, on one occasion the bike was parked outside overnight in a driving rain, and still fired on a single kick the next morning.

Triumph brakes get better and better. We noted in an earlier test report on the Thunderbird that the shudder has been removed from the front brake, and apparently this increased smoothness is not peculiar to one example, for this latest test Triumph was every bit as good. Actually, it cannot be said that the Triumph has a road-racing type front brake. The drum is entirely of iron, and entirely unventilated, so that whatever heat is created in stopping does not remove itself from the brake very soon. However, while the iron drum may not dissipate heat very quickly, it does not distort any either, and thus the braking is consistent and smooth, even though not up to the demands of repeated panic stops from 100 mph — circumstances that it will not have to cope with in any case. Interestingly, rear brake performance has been improved by the simple expedient of moving the brake actuating lever in closer to the backing plate. Previously, this was cranked out to allow the brake rod to pass outside the left rear suspension unit; now the actuating lever is in close and the brake rod is tucked in between the suspension strut and the wheel. This has eliminated the tendency for heavy pressure on the brake to warp the backing plate, and the rear brake now has a more solid feel. Things of this sort may seem small, but it is the sum total of such things, introduced over the years that the vertical twin Triumph has been in production, that has brought it to its present state of refinement.

Curiously, while there have been so many refinements in so many areas, the Triumph clutch still shows a reluctance to disengage completely, and is rather heavy in action. Not the heaviest, by any means, of any motor-

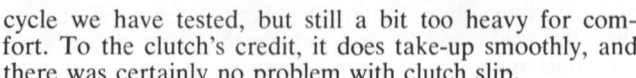

pace, cranking it down hard in the turns, and it never surges or wobbles.

The average rider will not push the Triumph hard enough to become aware of its potential as a road-racer; but he will appreciate the sure stability, and the soft, well-damped ride. This TR-6 SR Triumph also has a wide, soft seat, with enough length to carry two in relative comfort and, as before, the controls/seat relationship is "just right" for riders of widely varied size.

And then there is the good workmanship, and the wide availability of parts and service — which are considerations of some importance. Finally, there is the fact that the Triumph, especially this single-carburetor 40-incher, is one of the most pleasant of motorcycles to ride all day, every day. •

cycle we have tested, but still a bit too heavy for comfort. To the clutch's credit, it does take-up smoothly, and there was certainly no problem with clutch slip.

It is not our standard policy to take touring motorcycles to the race track, but circumstances brought this test machine to Riverside Raceway and we all had a fling at carving hot-laps. In doing this, we re-acquainted ourselves with the fact that Triumph engineers have worked a very considerable improvement in their motorcycle's handling. Of course, it cannot be said that the Triumph had bad handling before; it was quite a nice, stable touring motorcycle. However, the new forks introduced last year (and we have a feeling that the rear dampers may have been altered somewhat, too) have made the Triumph into a real road-racer. You can force it along at an insane

TRIUMPH TR-6 SR

SPECIFICATIONS

List Price	$1,169.00
Frame Type	tubular, single loop
Suspension, front	telescopic forks
Suspension, rear	swing-arm
Tire size, front	3.25-19
Tire size, rear	4.00-18
Engine type	vert. twin, ohv
Bore & Stroke	2.79 x 3.23
Displacement, cu. in.	39.6
Displacement, cu. cent.	649
Compression ratio	8.5:1
Bhp @ rpm	45 @ 6400
Carburetion	1⅛" Amal Monobloc
Ignition	battery and coil
Fuel capacity, gal.	3.75
Oil capacity, pts.	6.0
Oil system	dry sump
Starting system	kick, folding crank

POWER TRANSMISSION

Clutch Type	multi-disc, wet plate
Primary drive	duplex chain
Final drive	single-row chain
Gear ratio, overall:1	
4th	4.84
3rd	5.76
2nd	8.17
1st	11.8

DIMENSIONS, IN.

Wheelbase	55.5
Saddle height	31.5
Saddle width	10.2
Foot-peg height	13.0
Ground clearance	7.7
Curb weight, lbs.	396

PERFORMANCE

Maximum practical speed (after 1/2-mile run)	95
Max. speed in gears @ 6500 rpm	
4th	103
3rd	86
2nd	60
1st	40
Mph per 1000 rpm, top gear	15.8

SPEEDOMETER ERROR

30 mph indicated	actual 28.0
50	47.2
70	65.4

ACCELERATION

0-30 mph, sec.	2.2
0-40	3.1
0-50	5.1
0-60	7.0
0-70	10.3
0-80	13.7
0-90	19.2
0-100	28.0
Standing 1/4 mile	15.8
speed reached	85

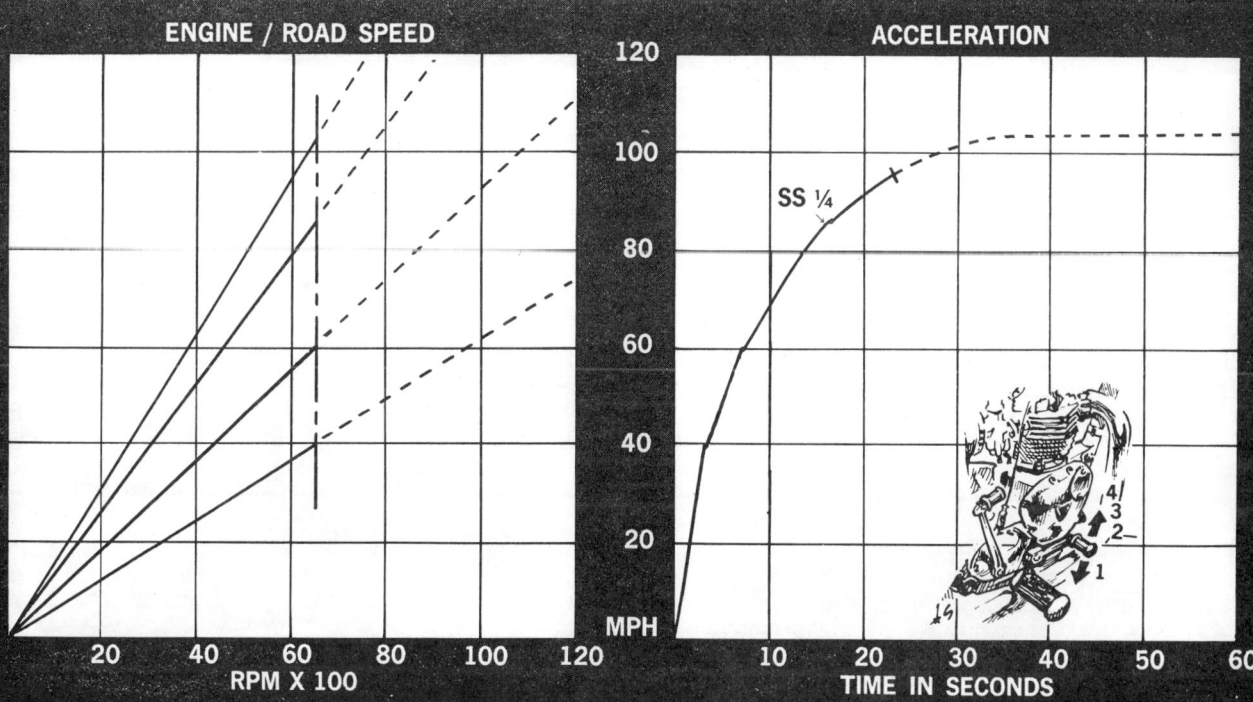

DICK RIOS'
TWO-TIMER

IN LESS THAN TEN SECONDS the Two-Timer accelerates to 155.17 mph, a mark which rider/builder Dick Rios claims as "the all-time mph (record) for a bike in the quarter mile." Rios also claims his 9.74 seconds elapsed time, recorded on the Fontana (Calif.) dragway clocks as "the lowest E.T. ever turned by a bike in the quarter." If anyone wishes to dispute these assertions, he can find Rios and Two-Timer at drag strips across the U.S. (and, hopefully, Europe) this summer on a coming exhibition tour. But whoever dares fling down whatever dragsters use for a gauntlet had better have his weapon well prepared, for without being overly stressed, Two-Timer lacks very little in the battle against the quarter mile. It consistently runs over 150 mph and Rios confidently predicts speeds in the '60s before the year is out, perhaps on his tour.

Power is generated by two 1957 650cc Triumph engines. Crankshaft and cases are stock, with .070 ground off the flywheel rims for cam clearance. Connecting rods are standard, with the exception of bronze inserts installed by Froyd & Weller of Pasadena. Harman & Collins cams are used with standard cam followers and lightweight pushrods made by Wert's Speed Shop of San Jose, Calif. Pistons are 11½ to one Triumph B range, set up at .008 clearance and .010 ring end gap.

Rios prefers to leave the cylinder barrels as standard as possible since they are more reliable that way, and he simply bores them out .010 at a time as they wear. If it begins to appear that the specifications are conservative, perhaps it should be mentioned that Two-Timer is a "fueler" designed to run reliably as well as quickly on up to 96% nitro, "which," as Rios says, "is almost the whole can."

Two-Timer's heads are aluminum, with optional 1 5/8″ intake valves, standard 1½″ exhaust valves, S&W valve springs with Webco alloy caps and keepers. Four remote float bowls feed four Amal 10TT9 carburetors with #7 slides coupled to a dual throttle. Standard magnetos deliver the spark.

Fitting the second engine in the 1950 frame required welding in 18½″ sections of tubing. The engines are joined in the center by alloy plates and standard engine mounts are used front and rear. Takasago 5/16x½ chain couples the engines through an ingenious set of splined sprockets which allows Rios to change gear ratios simply by slipping off the rear engine drive sprocket without disturbing the chain.

A Harley-Davidson clutch with Barnett plates handles the torque. Because of the pressure involved, the familiar Harley clutch booster is retained. Only second and third gears of the old H-D 3-speed transmission are used. While the hand shift lever is still there, the lower "clutch" lever discernible on the left handlebar is rigged to trigger the gearchange mechanism so speed shifts can be thrown without removing a hand from the bars.

The front forks are 1951 Triumph, fitted with sidecar springs and topped with 90 weight gear oil. Between the fork legs, where more prosaic motorcycles mount a headlight, Rios has installed a 10 lb. block of lead to help keep the front end down. Triumph brakes are used, with special linings on the front.

Two-Timer's oil tank once held three quarts of gasoline for a lawnmower, and the gas tank is an English item, purchased like most of the other parts from Harry Foster's Los Angeles Triumph shop, the closest thing to a sponsor Rios ever had.

To blast the Two-Timer through the quarter mile Rios pushes back the hand shift about ½-inch to engage second gear, winds the engines to peak torque rpm and skillfully slips in the clutch to turn the Avon racing slick at maximum bite without breaking loose. Formerly the "trick" was to release the clutch like a hot iron. If the power was there the slick would "light up" or break traction and become itself a clutch against the asphalt. Rios' new method, besides giving faster speeds, considerably extends the life expectancy of his tires.

Rios believes car dragster speeds (now well over 200 mph) can be equalled with "a motorcycle that still looks like a motorcycle." Certainly the Two-Timer fits that description to a double T. •

CYCLE WORLD
ROAD TEST

TRIUMPH TR6SC TROPHY SPECIAL

IN COMPETITION, a motorcycle will occasionally appear that stands head and shoulders above all others. Sometimes it will even hold this position for a full season, and if it should be dominant for more than one season it is virtually ready for the hall of fame. Now then, out in the trackless wastes of the California desert, the "hare and hound" and enduro events consist of riding furiously through mile after mile of impossible terrain. Clanging and banging over the rocks and blasting through the sand washes is indescribably hard on machinery, and it is nothing short of incredible that one motorcycle could win so many races and do it for so many years in succession. But, there is one motorcycle that does it, and that motorcycle is the "40-inch" Triumph.

This has been going on for so long that, at last, even the Triumph factory has taken note, and they now offer the TR6SC "Trophy Special," which is suitably outfitted for the kind of work we have described as it stands on the dealers showroom floor. The Trophy Special is, if you please, a real off-the-shelf fire-breather, and while it is entirely possible to get more-than-standard horsepower from its engine, that is not necessary. Despite the fact that it is outfitted with a hefty and very knobby rear tire, the rider can exceed available traction at any time, in any gear, and almost at any speed simply by turning on the power.

Naturally, with so much power the Triumph is a "born" slider, and the experts use this to good advantage. Instead of peeling-off into turns and going around all leaned over, the Triumph expert just tweaks the bike over in the direction of the turn and cranks on throttle. The amount of throttle controls the rate at which the rear wheel proceeds around and thus controls the rate of turning. If you have ever watched this sort of thing, you will have probably noted that all this action takes place with the motorcycle

51

nearly upright. It is the nearest thing to sprint-car driving that is possible on two wheels and it is only possible with thunderous power like the Triumph's.

Unfortunately, the Trophy Special is a close development from Triumph's touring bike, and a bit short of fork-rake for the speeds it will attain in a straight line over rough ground. For that reason, the devoted super-experts bend a few more degrees of rake into the steering head, heating the frame and applying a hydraulic jack. It is a trifle brutal, but it works and that is what wins races.

For the slightly less fanatical dirt-pounder, the Trophy Special will serve nicely as a dual-purpose racing/fun motorcycle. The long, soft seat inherited from Triumph's touring bikes makes it an attractive rig for hauling a passenger — even though no passenger pegs are fitted. Actually, a passenger can rest his/her feet on the high pipes, and the folding pegs provided for the rider are so long that there is almost room on them for everybody's feet.

Part of the TR6SC package is stiffer springs and long-travel forks, and these do an admirable job of ironing out bumps. The bike is comfortable for all concerned even when packing double, and while indulging in our favorite sport, jumping, we found that the suspension is virtually impossible to "bottom." More than one so-called cross-country motorcycle we have ridden has failed this part of the test — betraying its inadequacies with a ghastly metal-to-metal clank as its suspension hit the end of its travel.

Again, we will give thanks to the change to unit-construction for the engine/transmission. Purists moan about this, but it is a tidy arrangement and reduces the number of joints from which oil can leak. The Triumph, in some 600-miles of hard desert riding, stayed absolutely leak-free. Clean-up required only a light hosing off of accumulated dust and mud — except at the back fender, where the oil-tank breather hose had sputtered a generous amount of Mr. Castrol's best.

Those who might be tempted to replace the rather skinny-looking exhaust pipes with something larger in diameter should restrain themselves. The pipes are long and thin because that is what spreads the power around and makes the engine pull so strongly at low speeds.

One little item new to Triumph motorcycles is the top-center finding system. You remove a plug from the back of the crankcase and thread-in a substitute plug that is drilled to hold a sliding rod. This rod is held against the flywheel rim as the engine is cranked slowly over. When top-center is reached, the rod pops into a hole in the flywheel. Hopefully, the rod will be removed before the engine is started, for if it should drop home with the crank turning, serious damage will surely result.

All of the usual Triumph good things are provided on the Trophy Special. Not the least of these is the extreme ease of starting, which is really quite remarkable for a big-displacement twin. Also, there is the knife-through-butter gear change, complemented by a light, progressive clutch action. Finally, there is the finish, always good and getting better every year.

However, what we liked most about the Trophy Special is that it's such fantastic-fun to ride. We spent a lot of time pounding along the wheel-tracks of mountain access roads, and several hours out on a dry-lake bed, spinning huge, 70 mph "donuts" — laughing like maniacs all the while. The Triumph may not be the best-handling scrambler in the whole wide world, but it has more hair on its chest than King Kong and has proved time and again that it will simply out-muscle its better-handling opposition if its rider is similarly hairy-chested. The basic incompatability of knobby tires and pavement prevented us from proving all this at the drag strip; seat-of-the-pants evidence was enough for us out in the dirt. •

TRIUMPH TR-6 SC

SPECIFICATIONS

List Price	$1,145 f.o.b. POE
Frame Type	tubular, single-loop
Suspension, front	telescopic fork
Suspension, rear	swing arm
Tire size, front	3.50-19
Tire size, rear	4.00-18
Engine type	vert. twin, 4-stroke
Bore & Stroke	2.79 x 3.23
Displacement, cu. in.	39.6
Displacement, cu. cent.	649
Compression ratio	8.5:1
Bhp @ rpm	45 @ 6500
Carburetion	(1) 1⅛" Amal Monobloc
Ignition	energy transfer
Fuel capacity, gal.	3.75
Oil capacity, pts.	6.0
Oil System	dry sump
Starting system	kick, folding crank

POWER TRANSMISSION

Clutch Type	multi-disc, wet plate
Primary drive	duplex chain
Final drive	single-row chain
Gear ratio, overall:1	
4th	5.41
3rd	6.44
2nd	9.14
1st	13.5

DIMENSIONS, IN.

Wheelbase	55.5
Saddle height	31.5
Saddle width	10.5
Foot-peg height	12.6
Ground clearance	7.5 (at stand)
Curb weight, lbs.	379

PERFORMANCE

Top speed (at 7000 rpm)	99
Max. speed in gears @ 7000 rpm	
4th	99
3rd	83
2nd	58
1st	40
Mph per 1000 rpm, top gear	14.3

SPEEDOMETER ERROR

30 mph, actual	
50	(no speedometer)
70	

ACCELERATION

0-30 mph, sec.	1.6
0-40	2.5
0-50	4.0
0-60	6.1
0-70	8.3
0-80	11.3
0-90	16.2
0-100	
Standing ¼ mile	15.0
speed reached	88

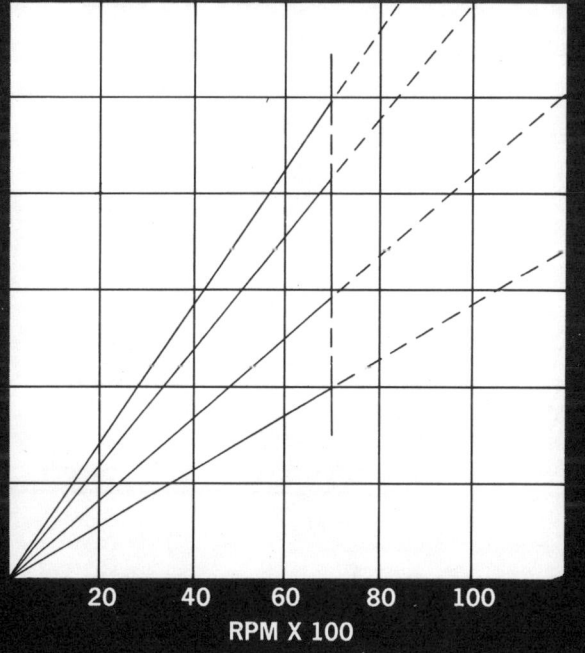

ENGINE / ROAD SPEED

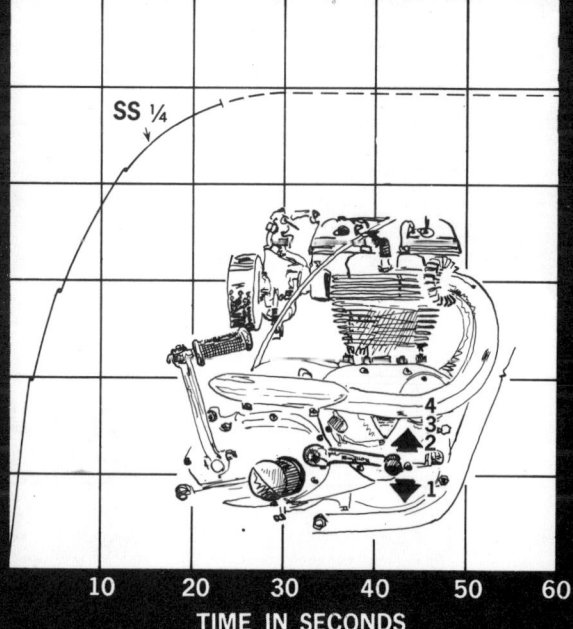

ACCELERATION

CYCLE WORLD
ROAD TEST

FOUR YEARS, AND 100 ROAD TESTS LATER, WE RETURN.
TRIUMPH T120/R BONNEVILLE

Cycle World's first issue, published four years ago, featured a road test report about Triumph's Bonneville. Forty-eight months and a hundred road tests later, we return to the Bonneville for our fourth anniversary issue. As one might expect, with the passing of so much time it has become a much better motorcycle, although there has been no major change and even though there is only a fractional increase in performance. The improvement has been in handling, braking and reliability.

Taking first things first (handling), this aspect has been improved by the addition of an extra 2 1/2-degrees of rake, which places the 1966 Triumph's steering head angle at 29 1/2-degrees from the vertical. This is rather more rake than one usually finds in a touring machine, and in fact there is a distinctly heavy feel to the steering of the new Triumph. Interestingly, the increased rake is not coupled with any increase in wheelbase, so evidently the steering head was moved back slightly when the angle was changed. This is, however, the extent of the changes in the frame, compared to 1965. The design is much like the duplex-cradle frame of 1962, but the tubes leading down from the steering head to the crankcase cradle have been replaced with a single, large-diameter tube. This single tube is neither stronger nor weaker than the pair of tubes it replaces but it resists transverse-plane vibrations better, and has the effect of making the entire motorcycle smoother.

Triumph claims that the "new" frame, with its added steering rake, was introduced to improve high-speed stability. Indisputably, it does what they claim — but there was nothing wrong with that aspect of the Triumph's handling in 1965. A much more likely reason for the change is found in the behavior of the big Triumph dirt-pounders in years past. These have always suffered from a tendency to pitch their riders off in fast, rough-ground going. The cure, applied by all the experienced scrambles riders, has been to bend more rake into the steering head. Now, Triumph has adopted this speed-merchant's setup for all B-range models. A good thing, we think. Touring riders will very likely prefer the added steadiness they will get with the new frame, and the racing bunch will have had one major modification removed from their list of things to do.

The Bonneville's brakes are better than before mostly because the front brake is larger. Not larger in diameter, it still has an 8-inch drum; but the drum has been made wider. The previous full-width brake was laced into the wheel with straight spokes, an arrangement with advantages in terms of strength but one that limits brake shoe width. The straight spokes require a row of holes around the mouth of the brakedrum, and the shoes must be set in deep enough to clear these holes. On the new brake, the spokes on the open side of the drum are laced to a flange. This makes room for wider shoes, and there has been a 44-percent increase in lining area. Also, the spoke flange around the mouth of the drum adds stiffness, so there is less tendency to fade during repeated hard stops from high speed.

No change has been made in the size of the rear brake, but there has been a big improvement made in the hub assembly. Last year, the big Triumphs had their rear sprocket as an integral part of the rear brake drum; the teeth being machined right into a flange cast as part of the drum. Now, the sprocket is a separate, bolt-on part, and changes in gearing can be made without adding a big, untidy "overlay" sprocket, or digging past the clutch to replace the countershaft sprocket. Actually, the overlay sprocket was not such a bad thing if you wanted to substantially increase the overall ratio; but there was no way to add, say, a half-dozen teeth to the rear sprocket.

Reliability has been improved in a variety of ways. In the engine, a change has been made from a ball-bearing to rollers on the drive-side main bearing, which is more of a change than one might think. Not long ago, Triumph Engineering was telling us how they had discovered that by pre-loading the ball-bearings, the engine would run much smoother. Unfortunately, there were also reports from owners that the pre-loading was shortening bearing life. The present combination of ball and roller main bearings is one long favored by Triumph-oriented tuners, and it impresses us as being an excellent arrangement.

Another small change with large implications is the provision of an oil feed direct to the exhaust tappets. Previously, both intake and exhaust camshafts relied on splash from the crank, but this lubrication was somewhat marginal for the exhaust cams. Due to the fact that the exhaust cams must push their valves open against a fairly considerable cylinder pressure, they are more heavily loaded and need the best possible lubrication.

Further improvements in overall lubrication have been added with a new, larger-capacity oil tank. The added capacity will reduce the oil's running temperature, but we suspect that the main reason for this change is because oil for the rear drive-chain is now taken from the engine's supply. Before, oil from the primary chain-case was sprinkled on the rear chain, but that system would sometimes empty the primary.

In addition to the things already mentioned, the Triumph 650cc engine also has a lightened flywheel (weight reduced 2.7 pounds) and new 9:1 pistons. And then there are "R" tappets and a "racing" inlet cam — used in combination with a "sports" exhaust camshaft. Pushrod-tube seals have been borrowed from the C-range engine, as these have proven better at keeping the oil inside the engine.

Having had no small amount of difficulty with the tachometer drive on our own Triumph Bonneville record-holder, we were much pleased to see that this item has been improved. In place of the cable emerging directly from the end of the camshaft, looping 'way out to the side before continuing on up the tachometer, the drive first goes into a right-angle drive box that reduces the cable drive to 1/4 engine speed. With the old arrangement, cables would kink, and break, and sometimes the housing would snag on something and be torn away. The new drive-box aims the cable along a straighter path, and tucks it in where it is less likely to have chance encounters with passing brush, etc.

The Bonneville's electricals have not been ignored in the general upgrading. Everything is now operated on 12-volts, with current drawn from a pair of 6-volt batteries linked in series. The batteries are charged by the usual crankshaft-mounted alternator, which has a permanent-magnet rotor and fixed generating coils — all housed inside the primary case. The charging rate is controlled by a Zener diode, a "solid-state" device that progressively reduces output to the battery as it reaches full capacity. The Zener diode is a very superior output regulating device, and generally quite reliable. However, heat, vibration, voltage surges and sometimes even the phases of the moon can make them suddenly decide not to do whatever it is their little molecules are doing. A spare, tucked away with the spare spark plugs, would be a good thing for the Bonneville owner to have at hand. The Zener diode is a better device than any alternate means of voltage regulations, but when it packs up, it cannot be coaxed back into action.

At last, the Triumph factory has heeded the pleas of their American distributors and are sending a motorcycle with a fuel tank not cluttered by a parcel grid. The small carrying grid found on earlier Triumphs is doubtless a handy thing, but Americans have never cared much for it. We view its passing as a sign that the Triumph factory is beginning to care more for Americans.

Again, we will say that we very much like the Triumph Bonneville, and this latest model is the best yet. The stainless-steel fenders are a practical touch that present a fetching appearance, as does the new tank with its "racing-stripe." Perhaps because of the lighter flywheel, or perhaps because the handlebars are rubber-mounted, the bike seems even smoother than before. Triumphs have always had quite a good riding position and controls layout, and the new Bonneville is no exception. In all, the Triumph Bonneville for 1966 must still be considered the standard by which all medium-large displacement, high-performance touring bikes are judged. Others may be faster, *or* better handling, *or* even slightly more reliable. None offers quite the same combined package of performance, *and* reliability *and* handling. ∎

TRIUMPH T120/R BONNEVILLE

SPECIFICATIONS

List price	$1309
Suspension, front	telescopic forks
Suspension, rear	swing arm
Tire, front	3.25-19 ribbed
Tire, rear	4.00-18 block
Brake, front	8 x 1.625
Brake, rear	7 x 1.187
Total brake swept area, sq-in	66.5
Brake loading (test weight/swept area) lb/sq-in	8.16
Engine type	vertical twin, ohv
Bore and stroke (inches-millimeters)	2.79 x 3.23-71 x 82
Displacement (inches^3-centimeters^3)	39.6-649
Compression ratio	9.0:1
Carburetion	(2) 1 3/16" Amal Monobloc
Ignition	Lucas battery and coil
Bhp @ rpm	52 @ 6500
Oil system	dry sump
Oil capacity, pts.	6.0
Fuel capacity, gal.	2.5
Starting system	folding-crank, right-side
Lighting system	Lucas AC, 2 x 6V. batteries
Air filtration	paper-element
Clutch	multi-disc, oil bath
Primary drive	duplex chain, clutch-hub cushion
Final drive	single-row chain
Gear ratios, overall:1	
5th	none
4th	4.84
3rd	5.76
2nd	8.17
1st	11.8
Wheelbase	55.2
Seat height	31.5
Seat width	10.2
Foot-peg height	13.0
Ground clearance	7.1
Curb weight (w/half-tank fuel)	392
Test weight (fuel and rider)	542

PERFORMANCE

Top speed	109
Maximum speeds in gears (@ 7000 rpm)	
5th	none
4th	111
3rd	92
2nd	64
1st	43
Mph per 1000 rpm, top gear	15.8
Speedometer error	
30 mph indicated, actually	28.2
50	47.7
70	66.2
Acceleration, zero to -	
30 mph, sec.	1.8
40	2.4
50	4.0
60	5.7
70	8.3
80	11.2
90	14.9
100	21.0
Standing 1/4-mile, sec.	14.2
terminal speed	88

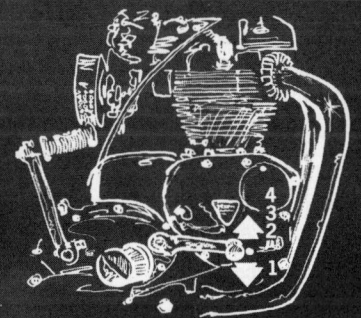

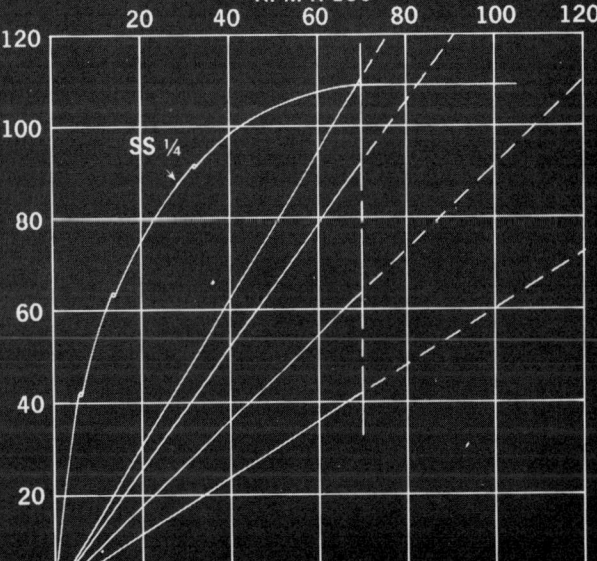

TRIUMPH HISTORY

BY GEOFFREY WOOD

THE TRIUMPH MOTORCYCLE, like many of its other British brothers, had its beginnings in the bicycle industry. Founded in 1885 in London, the early history of the Triumph Company is very nearly lost in the mists of time.

The founder of the company was, surprisingly, not an Englishman, but rather a German by the name of Siegfried Bettmann. Bettmann's small bicycle manufacturing business prospered, and in 1887 he was joined by a fellow German, M. J. Shulte, who was a young design engineer. Together, these two Germans were later to have a profound effect on the infant motorcycle industry.

In 1888 the company moved to new quarters in Coventry, and in 1897 the brilliant Shulte was investigating the possibilities of a motor-powered bicycle. The motor-bicycle under examination was the Hildebrand and Wolfmuller, a German machine which Shulte regarded as being far too crude to ever be a workable design. The seeds of a motor-powered bicycle were sown in Shulte's mind, though, and he was convinced that the future held great things for a motorbike.

It was in 1902 that the first motorcycle was produced by the burgeoning company, when a Belgian Minerva single-cylinder engine was mounted on one of their bicycles. The original 66mm bore by 70mm stroke engine had an automatic inlet valve, battery-coil ignition, and it was mounted below the front downtube of the bicycle. For 1903 the engine was modified to standard side-valve design, and in 1904 the Triumph had a British J.A. Prestwich engine which was similar to the Belgian Minerva powerplant. The factory also produced a model that year with a larger 3 hp Belgian-made Fafnir engine mounted centrally in the frame. All these early Triumphs had belt drive and bicycle pedal-gear with a single rim-brake.

Despite the early enthusiasm for the motorbike, the public had found it lacking in reliability and overall performance. With this in mind, the brilliant Shulte designed the very first all-British motorcycle in 1904 and production began the following year. The engine was a 3 hp single-cylinder side-valve, centrally mounted in a properly designed motorcycle frame. Ignition was by a reliable magneto, and the carburetor was of their own design.

To publicize this first all-British machine the company staged a demonstration run to prove its endurance. The goal was to cover 200 miles per day for six days — a really arduous test for machines of that era. The run was a success, and the new Triumph was on its way to tremendous popularity.

The 200cc Cub model in genuine trials trim. First lightweight ever to win the famed Scottish Six Days Trial, the Cub has established a great reputation in European trials events. The same model with lights is marketed as the Mountain Cub in the U.S.

Production in 1905 was at the astounding rate of five machines per week, which rose to a healthy 500 machines for 1906. It was in 1906 that a front fork with a suspension spring was introduced, which made for a more comfortable ride. In 1907 the engine dimensions were enlarged to 82mm x 86mm and production increased to 1000 machines per year. In 1908 the engine was again increased to 85mm x 88mm, which made it a full 500cc; and in 1909 the production was up to 3000 units.

During those very early days Triumph became renowned on the race courses, and their trusty little singles achieved many victories. In the very first Isle of Man TT race, for instance, the marque garnered second and third places in the single-cylinder class. In 1908, Jack Marshall copped first place at 40.4 mph with a fastest lap at 42.48 mph. Other Triumphs took third, fourth and fifth places. The models featured direct belt drive at a ratio of 4.5 to 1.

A clutch in the rear hub was introduced in the 1911 models along with such niceties as adjustable tappets to set valve clearances. In 1913 a three-speed Sturmey-Archer hub gear was available, and a 225cc two-stroke was also added to the range. The company experimented with a side-valve vertical-twin engine, too, but it never did reach the production stage. This last item was a hint of what was to come in later years.

For 1914 the bore and stroke dimensions

The original 1938 Speed Twin marked the beginning of the vertical twin's popularity. Girder front fork and rigid frame were standard for their day. Top speed — 90 mph.

PIONEERS OF MOTORCYCLE DESIGN SINCE THE TURN OF THE CENTURY

were increased to 85mm by 97mm, making it a 547cc engine. For the sporting rider a 500cc model was still available, though, as the international racing regulations limited engines to that size. The reliable Triumph single continued to make its mark in competition, such as W. F. Newsome's one-hour endurance record at 59.84 mph in 1910. The famous Isle of Man TT races continued to witness the reliability of the side-valve single as the marque garnered third in 1909, third and fourth in 1910, sixth in 1911 plus the one-hour record at 63.11 by J. R. Haswell; second, fifth and sixth in 1912; and a fifth in 1914.

Then war settled over Europe and the factory became mobilized for military production. It was, in fact, in 1915 that the model H, which proved to be such a sound design, made its debut. The 547cc side-valve engine operated through a three-speed Sturmey-Archer countershaft gearbox with a final drive by belt. Altogether about 30,000 model Hs were built for the British Army.

After the war the model H was produced for civilian use, and this was superceded in 1920 by the S.D. — a model with all chain drive and a three-speed gearbox of their own manufacture. Then chief design engineer Shulte retired and Mr. Bettmann brought in Lt. Col. C. V. Holbrook, C.B.E., to replace him. The next model to come out of the factory was in 1924 when the 350cc model L. S. was introduced which had such advanced features as unit construction and full mechanical lubrication. Previously, the lubrication was provided by a hand or foot-operated oil pump.

During the early 1920s the British switched from side-valve engines to the overhead valve design, and Triumph was right to the forefront. In 1922 the factory entered a four-valve penthouse-head engine designed by Sir Henry Ricardo in the Senior TT, and Walter Brandish took second place with it. The model had a "cast-iron" engine with a bore and stroke of 85mm x 88mm, and it was one of the very first uses of the now popular slipper-skirt piston. The model also hoisted the classic hour record to 76.74 mph in the hands of Major H. B. Halford.

For 1923 the factory raced a two-valve 494cc single with measurements of 80.5 x 98mm and scored some significant victories. Probably the most noteworthy was Victor Horsman's one-hour record of 86.52 mph. The following year he raised this to 88.21, and in May of 1925 he pushed it on up to 89.13 mph. Then he lost the record; and so in October he again made an attempt, raising the speed to 90.79 mph. This last mark was the first time a 500cc engine had pressed 90 miles into one hour. Then in 1926, Victor rode his faithful old single to a 94.15 mph mark, and that proved to be his swan-song.

During the middle 1920s the international racing scene was undergoing a great change. Previously, the machines were basically standard production models suitably prepared for racing, but by 1926 it had become necessary to build a genuine racing machine if there was to be any chance of success. It was then, and still is, the policy of Triumph to race only what they sell; and so when it became evident that a "tuned" standard model had little chance of success, the interest in racing justifiably diminished. Tommy Simister did take the ohv single into a third place in the 1927 Senior TT, but other than that the name of Triumph faded away from the great racing courses.

All the knowledge gained from racing and record setting was not lost, though, and in the late thirties the company generally switched over to ohv engines. The 500cc Model P made its debut in 1925 at the modest price of 41 pounds 17 shillings (about $117) and 28,000 were produced the first year. A sporting edition called the model Q was also available in 1927, and a 277cc side-valve lightweight model W was produced.

In 1928 Triumph changed their historic grey with olive-green fuel tank colors to black with blue panels. In 1929 a 350cc ohv model was added to the range, and

In 1903, Triumph featured a Belgian Minerva side-valve engine mounted on the standard bicycle frame. Pedal gear and clutchless belt drive were accepted practice in early days.

The 1914 side-valve Triumph single helped establish the early reputation of the company. A reliable mount for its day, the machine had belt drive and wheel-rim brakes.

dry sump lubrication was adopted. In line with the general trend in the motorcycle industry, a shorter wheelbase tubular cradle-frame was used, and the fuel tank was the "saddle" type that hung over the top frame tubes instead of between them.

Then came the depression of the early 1930s and sales fell off. The whole range of Triumph machines was reorganized by A. A. Sykes and a small, cheap two-stroke was produced along with some big singles of "sloper" design. By 1934 the company was making great strides in its automobile manufacturing, and Val Page responded by designing new vertical singles of 250cc, 350cc and 500cc. A 650cc vertical twin was also produced which had even firing intervals with the pistons rising and falling together, unit construction, and a geared primary drive.

It was common knowledge by 1936 that the company was not doing well, though, and that their motorcycle production was going to be halted. Soon the Triumph would be a memory of the past — the only models visible would be preserved in the museums.

Just at that darkest hour a man by the name of J. Y. Sangster came forward to purchase the company, and the concern's new name became the Triumph Engineering Co. Ltd. Sangster had previously been with the Ariel Company, and one of his first moves was to appoint Edward Turner as chief design engineer. Turner had done a great amount of design work on the Ariel Square Four and Red Hunter single — and these models had proved eminently successful. Would this be enough to save the Triumph?

The impact of the Turner genius on the new company was immediate and positive. A new range of single-cylinder machines was marketed in 1937 that acquired a reputation of being good looking, reliable and brisk performing. With these new thumpers, sales rapidly climbed and the future of the company was assured.

For the 1938 range another new model made its debut, and this machine can be said to have redesigned half of motorcycledom. Called the Speed Twin, the new 500cc model featured a very compact vertical twin engine with the pistons rising and falling together but firing alternately. With the addition of the twin to the range, Triumph had a mount for just about every use except racing.

The lowest priced were the 2H and 2HC models, which were 250cc ohv singles that developed 13 hp at 5,200 rpm. The compression ratio was 6.9 to 1, and the rigid frame bike weighed 310 pounds. Both of the lightweights were popular machines.

Next were the 350cc singles, two side-valvers and one ohv model which developed 12 hp at 4,800 rpm and 17 hp at 5,200 rpm respectively. Then there were the Deluxe 5H and Deluxe 6S models in 500cc and 597cc. The 5H ohv model developed 23 hp at 5,000 rpm on a 5 to 1 compression ratio, and the 6S side-valve model produced 18 hp at 4,800 rpm. The measurements of the two engines were 84 x 89mm and 84 x 108mm. The 360-pound 6S model proved popular for sidecar use.

The pride of the range was, no doubt, the single-cylinder "Tiger" model. Produced in 250cc, 350cc and 500cc sizes, the Tiger models were "sports" machines which gave an improved performance over the standard machines. The Tiger 70 produced 16 hp at 5,800 rpm on a 7.7 to 1 compression ratio and weighed 310 pounds; the Tiger 80 developed 20 hp at 5,700 on a 7.5 ratio and weighed 320 pounds; and the Tiger 90 had 28.3 hp at 5,800 rpm on a 7.1 ratio and weighed 365 pounds. These were the 250, 350 and 500 respectively.

The star of the Tiger range was the 500cc model, and it had husky 7" x 1-1/8" brakes, narrow fenders, a 20" x 3" front tire, and a 19" x 3.50" rear tire. All the Tiger models could be ordered in competition trim, which made them suitable for trials or scrambles use. Optional extras included an upswept exhaust pipe, knobby tires, a wide ratio gearbox with ratios of 4.78, 6.93, 11.00, and 14.7 to 1, quickly detachable lights and rear wheel, and individually assembled and tuned engines. As on all Triumphs the frame was rigid and the front fork was of a girder type. The Tigers were altogether a fine mount for the sportsman, and all this at the modest price of 77 pounds ($385).

Then, of course, there was the new Speed Twin which had measurements of 63 x 80mm and produced 28.5 hp at 6000 rpm on a 7 to 1 compression ratio.

The 1938 Tiger 90, a 500cc model in competition trim. These pre-war singles had an excellent reputation for reliability — were popular with trials and scrambles riders.

The 1949 Grand Prix model in all its glory. A potent racing mount with sleek lines, the GP provided private riders with a speedy machine that cost well below an overhead camshaft model. Note the massive front brake, spring hub, and twin carbs with remote float.

The twin weighed 365 pounds, had a 54" wheelbase and sold for 75 pounds ($375).

This new range of Triumph machines proved very popular, and the sales of the company continued to expand. The sports singles won their share of the trials and scrambles events, and the Speed Twin set an all-time Brooklands 500cc lap record at 118.02 mph. The twin was built and ridden by I. B. Wickstead, and the model featured a supercharged engine. Another Brooklands all-time lap record was also garnered by Triumph, this one in 1939 by W. F. S. Clarke at 105.97 mph on his 350cc fuel burning single.

For 1939 the range of machines and optional parts offered stayed the same as for 1938 except that an exciting new Tiger 100 replaced the Tiger 90 model. Based upon the Speed Twin, the new Tiger 100 was a mount to satisfy the most discriminating buyer in 1939. The engine was individually built, tuned, and tested on a dynamometer; and it was certified to produce 33 to 34 hp at 7,000 rpm on a 7.75 to 1 compression ratio. Each owner received an actual dynamometer report with his engine, signed by the works test mechanic.

The rest of the specifications whetted the appetite of the motorcycle connoisseur. The mufflers were designed to be megaphones with end caps, and by removing these caps an owner was ready to race. The front wheel had a 3.00" x 20" ribbed tire, and the rear wheel had a 3.50" x 19" semi-racing tire. The front brake drum was ribbed for extra cooling, and both brakes were 7" x 1-1/8" in size. A special bronze head was optionally available for the Tiger 100 which gave improved heat dissipation, and a set of tuned straight pipes were also available. With a top gear ratio of 5.0 to 1, the 7000 revs with the open megaphones gave a top speed of 106 mph. Truly, this was the machine that the British motorsportsman had dreamed of.

Then came the second war, and on a dark November night the factory was left a pile of twisted beams and smoldering rubble. To this day no one knows quite how it was done, but in ten months a new factory was built at a new location in Allesley, near Meriden. And there during the war, Triumph produced a 350cc ohv single for military use.

When peace returned Triumph immediately resumed production, and a new range of machines was fielded. Gone were the single-cylinder models, and Triumph began production of twins exclusively. A new telescopic front fork replaced the old girder fork, and in 1947 the Triumph "Spring Hub" made its debut. This Spring Hub was a rather unique method of rear suspension as it contained the coil springs within the large rear hub.

A new 350cc twin was also added to the range to supplement the Speed Twin and Tiger 100 models. In deference to the limited amount of cash available in those immediate post-war years, the Tiger 100 was not quite so sporting a mount as in the pre-war days. Horsepower was down from 34 at 7000 to 30 at 6,500 rpm, and all the hand assembled and dynamometer tested qualities were gone. The brakes were not finned, no megaphone mufflers were fitted, and few of the pre-war optional "goodies" were available.

The basic design was still very good, though, and quite naturally some sporting-minded riders turned their attention to tuning the Tiger 100 for racing. Many racing men recognized that a small-bore twin had a great advantage over a big-bore single as far as getting the highest possible compression ratio on the dreadful 72 octane "pool" petrol that was used.

It was Ernie Lyons, the Irish farmer-racer, who got the show on the road. Working with factory support, Lyons took a standard Tiger 100 and began building his racer. During the war Triumph had made an electrical generating power-unit for Lancaster bombers using a standard 500cc twin engine which had an alloy head and cylinder with fan cooling. Ernie borrowed this alloy head and cylinder part and also obtained the experimental Spring Hub to use. With just the normal amount of speed tuning for the day, Ernie was ready to compete.

The first event for the racing Tiger was the 1946 Ulster Grand Prix. All sorts of problems were encountered that day, but the bike did show some dazzling, if rather spotty, performance. Nevertheless, Lyons was encouraged and he set about to cure all the "bugs." In September he and his Tiger appeared again, this time at the Manx Grand Prix for amateurs held over the famous Isle of Man TT course. All went well that day, despite the appalling rain, and Ernie romped home the winner at 76.73 mph.

About that time the folks at Meriden began to give some serious thought to this racing game, and so for the 1947 season a prototype racer was prepared for the Grand Prix racing season. The late David Whitworth, a top-flight racing man, was engaged to ride; and he spent the summer touring the continental events. Whitworth had a highly successful season, too, winning minor Grand Prix events at the Circuits de La Cote, du Limbourg, and George Truffant. David also captured a third place in the Dutch Grand Prix, beaten only by the factory Norton of Artie Bell and the Gilera "Saturno" single of O. Clemencich.

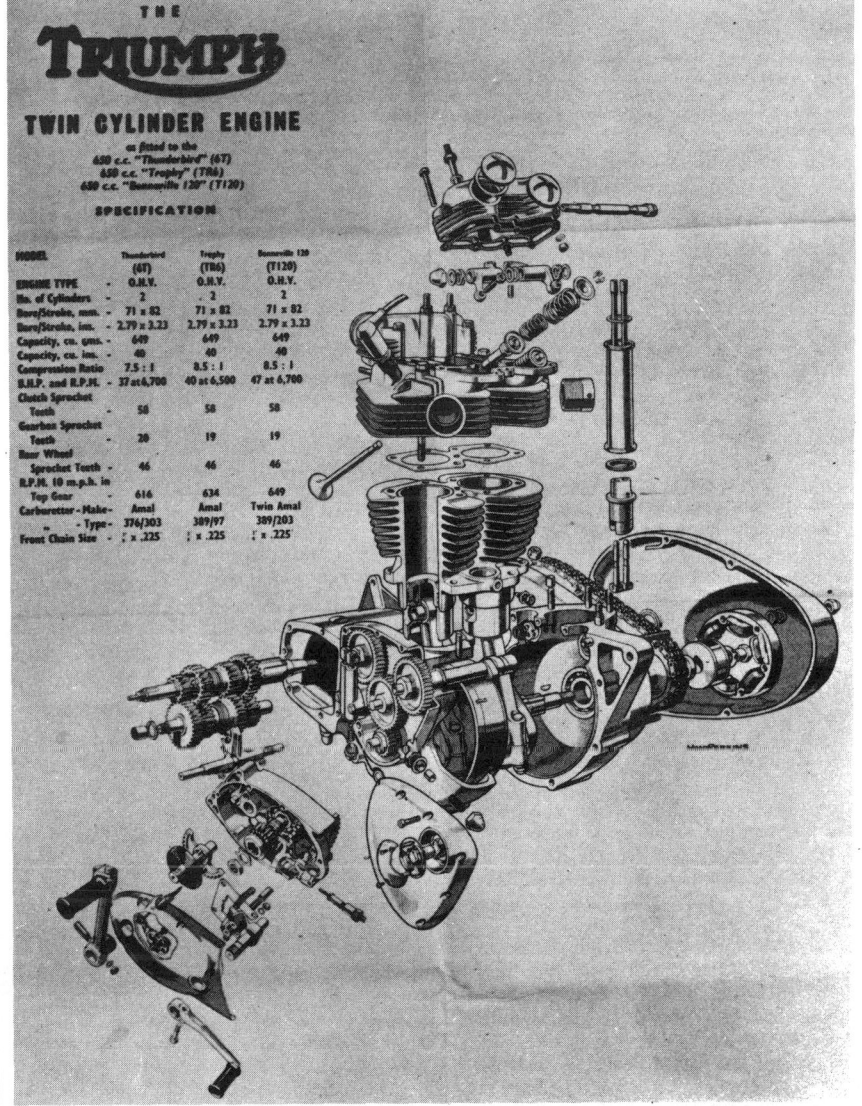

Engine internals of the 650cc Triumph engine that has proven so successful. High location of the camshafts gives short, light pushrods — a real asset for high rpm. The AC generator runs off the opposite side of the crankshaft, another Triumph advancement.

TRIUMPH HISTORY

Encouraged by these successes, in late 1947 the factory announced that a production road racing model would be marketed for the 1948 season. Called the "Grand Prix," the racer incorporated all the knowledge that had been gained during Whitworth's campaign. The idea was certainly not to build the best racing machine available, but rather to use as many existing standard parts as was possible in an effort to keep the price down. In this manner a beginner could have a pukka racer at a price well below that of an overhead camshaft racer, and yet still have a speedy and reliable mount.

The new Grand Prix model was a beautiful machine, and it was fast. The engine was a 500cc twin with standard bore and stroke of 63 x 80mm, and an alloy cylinder and head were fitted. Megaphone exhausts were used, the standard frame had the rear spring hub, and both brakes were a massive 8" x 1-3/8". The wheel rims and fenders were of light alloy, with a 3.00" x 20" ribbed racing front tire and a 3.50" x 19" studded racing rear tire.

Each engine was mirror-polished throughout, and the cams were designed to give a great amount of torque over a wide rpm range. Two Amal carburetors were used, and the float bowl was remote mounted. The engines were all hand assembled and tested on a dynamometer. On a 10.5 to 1 compression ratio the horsepower was 42 at 6800 rpm. Top gear of the four-speed close-ratio box was 4.57 to 1, and at 6800 revs this provided a maximum speed of 112 mph. On any downhill run the revs could safely soar to 7400 rpm, which gave a speed in excess of 120 mph. Another good point in the GP model's favor was its 314-lb. weight, which was well below the 370-lb. weight of a Norton Manx model.

So enthused were the Triumph folk over the new racer that they decided to set aside their policy of no direct participation in racing. In short, they fielded a genuine works team mounted on some GPs that received that little extra bit of tuning and bearing work that always helps. They also hired some of the finest riders available with such famous former TT course winners in the lineup as Bob Foster, Fred Frith and Ken Bills.

Altogether a total of nine Grand Prix Triumphs were entered in the Senior TT — but then followed a tale of disaster. One by one the riders fell by the wayside — gas tanks fractured, gearboxes disintegrated, clutches failed and engines blew. In the end, not one GP model finished. Shaken but not yet defeated, the designers went back to the race shop and produced a new fuel tank and made minor changes in the other engine and transmission parts that had failed.

On the continent the small improvements revealed at last the excellence of the basic design. In the Dutch Grand Prix, David Whitworth took fourth; in the Ulster, Bill Beevers took fifth; and in the fast Belgian event, Foster, Whitworth and Bills took a magnificent fourth-fifth-sixth. Then, in September, Don Crossley won the Manx Grand Prix for amateurs in the I.O.M. at a post-war record speed of 80.63.

Despite the late season successes, the memory of the embarrassing Isle of Man episode brought the decision not to participate in racing anymore. The factory did produce the GP model for one more year, though, and many loyal privateers carried on with their twins. In the 1949 Senior TT they had a truly great hour, too, with Sid Jensen of New Zealand and C. A. Stevens taking fifth and sixth places, the former at 86.928 mph. These two Triumphs were the first non-works bikes to finish — a really magnificent achievement.

But then Triumph faded away from the racing scene and concentrated instead on improving their standard production machines. In 1950 the now famous Thunderbird model made its debut with a husky 650cc engine that had measurements of 71mm x 82mm. Then there followed the alloy-engined Tiger 100 and Trophy models, the latter establishing a great reputation in the trials world. It was the Thunderbird that once again set a trend, though, just as the original Speed Twin did in 1937. With its beefy power it has proven exceptionally popular all over the world, and particularly so in the U.S.

Never ones to sit on their laurels, the design improvement work went steadily on. In 1954 a new swinging arm frame made its debut, and also a snarling 42 hp model called the Tiger 110. Then there was the 150cc Terrier single, a light and inexpensive mount that was later developed into a 200cc model. Then, in 1957, came the new 21 cubic inch model that featured comprehensive enclosure and provided a "gentleman's" machine.

Today the Triumph range is even greater. The 200cc Cub can be had in road or genuine trials trim, this latter bike being the first lightweight to ever win the famed Scottish Six Days Trial when Roy Peplow did the trick in 1959. Based on the popular trials model, the Mountain Cub is built especially for the U.S. market and features lighting equipment and a 3.00" x 19" front tire instead of the 2.75" x 21" size.

Then there are the two "short-stroke" 500cc models with bore and stroke of 69 x 65.5mm, one in a street trim and the other in sports attire. If it's power you want there are the 650cc models, the standard Thunderbird and the potent Bonneville. The Bonneville is one of the fastest machines going today, and this twin-carburetored bomb has many victories to its credit in long distance European production-machine racing.

Produced especially for the U.S. market also, the TR-6 and TT Bonneville models are competition mounts for cross-country and TT racing. Their record in U.S. competition events is so great that it suffices to state that they simply dominate these two sports. The 350cc twin and the new scooter round out the range, although these models are vastly more popular in Europe than in the U.S.

And so this is the story of Triumph — a story of progress and achievement. Even today they are not sitting still, though, as stories keep filtering out of England about a three-cylinder 750cc model and an overhead cam 250cc Cub. One can just not imagine Triumph taking a back seat to anyone — history teaches us differently.

∎

The TRIUMPH line for '66 has

Step up to pleasure. It's the biggest year yet in motorcycling. Thousands of new riders are turning to the two-wheeled sport in a search for the real thing in thrills. And now, hard on the heels of the most successful year in history, Triumph again sets the pace with a daring new line for '66. A sleek, modern look from gas tank to tail light. A bold new thrust of power in the famous OHV Triumph engine. And a new high in performance, unequaled anywhere.

There's a bold new Triumph for every rider ready to take the first step to a world of new excitement. Take the first step up — to Triumph. Get the model designed for you. See your dealer now.

Fun means going places, and that's what Triumph owners do. You'll see them everywhere.

Big-bike economy, rugged construction and years-ahead styling set off the new 66's.

Note: Western specifications may vary from those shown.

T120/R — Bonneville Road Sports
Twin Carburetors. 40 cu. in. (650 c.c.)

The fastest standard motorcycle in the world today. Increased power and performance with unitized engine-gearbox construction. For the expert rider. ■ Color: New Alaskan White with Grenadier Red racing stripe. Polished stainless steel fenders.

TR6/R — Trophy Road Sports
Single Carburetor. 40 cu. in. (650 c.c.)

Long distance favorite of road sports riders. The top performing single carburetor model. ■ Color: New Two-tone Pacific Blue and Alaskan White.

6T — Thunderbird
Single Carburetor. 40 cu. in. (650 c.c.)

Completely re-designed with a brand new look. For Road Touring. Easy starting, economical, reliable. ■ Color: New Two-tone Grenadier Red and Alaskan White.

T100/C — Triumph Competition Tiger
30.5 cu. in. (500 c.c.)

The famous Jack Pine Model. Consistent winner in National Enduro events. ■ Color: New Sherbourne Green with Alaskan White racing stripe.

a model designed for you.

T120/TT
Bonneville Special Competition
Twin Carburetors. 40 cu. in. (650 c.c.)

For the competition expert who wants top performance. Designed for off the road racing and competition use only. ■ Color: New Alaskan White with Grenadier Red racing stripe.

TR6/C
Trophy Competition
Single Carburetor. 40 cu. in. (650 c.c.)

Patterned after the highly successful Jack Pine Model. Top reliability and performance. For woods or trail riding. ■ Color: New Pacific Blue with Alaskan White racing stripe.

T100/R
Tiger Road Sports
30.5 cu. in. (500 c.c.)

Truly a tiger. Versatile "snappy" performer. ■ Color: New Two-tone Sherbourne Green and Alaskan White.

T20/M
Tiger Mountain Cub
200 c.c. Lightweight

Has "Big bike" ruggedness for toughest woods and trails. 4 cycle OHV. Single Cylinder. Choice of experts and novices alike. ■ Color: New Two-tone Grenadier Red and Alaskan White.

As the fun goes, so goes Triumph. From beach to mountains, you'll find Triumphs being ridden by active people.

For the practical set, a Triumph's versatility hits the spot. At home flat out or on the green.

Who wants a TRIUMPH most?

Active people ride Triumph. And no wonder. A Triumph is built to stand up to tough riding. Mile after mile, year after year. And now, in 1966, Triumph presents a new look of action in a full range of models. Each new model has the big-pleasure surge of power that people on-the-go expect. Do you like the beach? There's more fun when you get there on your Triumph. Are you a cross-country fan? Those hills will flatten fast when your Triumph takes over. And if competition's your meat, do as the experts do — choose a Triumph! Who wants a Triumph most? Take a test ride. The answer will be obvious.

Triumph — The World's Best and Fastest Motorcycle. Triumph holds the A.M.A. approved WORLD'S ABSOLUTE SPEED RECORD, Bonneville, Utah, 230.269 m.p.h. (with streamlined shell).

Johnson Motors, Inc.
WEST...
P. O. Box 457. Pasadena. Calif. 91102

The TRIUMPH Corporation
EAST...
Baltimore, Maryland 21204

Write for FREE catalog in full color and name of dealer nearest you. Dept. K

Mounting brackets for extra Girling shocks are visible on swing arm, system proved unnecessary, providing too much dampening, according to Mulder.

Double spark plug leads for emergency ignition, allowing a switch to total loss battery system, coils for which are mounted behind seat. (PS: Never used).

Author Pat Owens, and stars of this tale, Eddie Mulder and his Triumph.

Eddie Mulder's winning mount.

BY PAT OWENS

A MOST VERSATILE MOTORCYCLE

FOR YEARS motorcycle racing in the United States has been confined for the most part to standard production machines, of the type generally available to the public. This is the core of the American Motorcycle Association's Class C rule, it has not been exactly inviolate. In most cases, the successful machines were built for one specific type of competition, most remain in their idiom until the end of their usefulness or until they are worn out. Some are converted back to street use.

Three years ago, California-based Triumph tuner Pat Owens started a project that was to prove just how many kinds of racing one motorcycle could be prepared for, and enter successfully. Starting point was a 1963 Triumph TR-6C which came from Meriden, England without headlights and equipped with an A/C magneto ignition system. Work began in March so a deadline was facing him; the Hare and Hound National about six weeks off. Time being a problem, only major alterations were made initially; changing the frame head angle, enlarging oil tank capacity, special foot rests, skid plate, a large capacity air cleaner, and some reworking of the forks.

Here's how Pat tells the story:

"As with most similar projects, work was finished and the machine first fired

The sensational Mulder in action on the California desert.

Late style front wheel mounted in forks backward so early style wheels can be used if needed, without changing right fork leg and fender braces.

54 tooth overlay sprocket gives 5.44 gearing for desert and drags, removed it gives 4.84 for street, with 19 tooth countershaft.

VERSATILE MOTORCYCLE

Triumph twin engine is sprayed with heat resistant black paint to assist cooling.

up the day before the event. Eddie Mulder, who had won the same event the year before on another machine I built and tuned, was to ride. He took his first ride only 18 hours before the event started. 'Midnight oil,' the racer's lubricant, was burned at a furious rate, and Mulder wound up with only ten minutes to practice before the start.

"Eddie got off to a fair start, ran tenth, then took the lead. At the end of the first loop he had more than a minute lead even after stopping for gas. He won in the rain with four minutes to spare. This proved to be just the beginning of racing successes as Eddie won the California State Hare & Hound Championship that season.

"Also in 1963, Al Rogers rode the bike in the Jack Pine enduro, finishing 9th Expert in the 'A' heavyweight class Rogers could have done better in this Eastern event. But his $1.50 pocket watch failed him on the morning of the first day. He arrived at a check nine minutes *early,* but finished the 500 miles without problems.

"In the spring of 1964 the machine was put back into street trim, and furnished my daily transportation. During a two-week summer vacation I rode it back to my ex-home in Connecticut, covering over 700 miles a day to save time for visiting. Converting the Triumph back to street use was actually fairly simple as the alternator which operates the A/C magneto also has built-in lighting coils. Gear ratios and the exhaust system are esily changed for road cruising. At the end of the 6,000-mile trip the motorcycle took on a new look with number plates, Class C traction tires and Ascot-type exhaust pipes, standard on the TT Special Triumph Bonneville."

The annual Corriganville Grand Prix was its first event, on the 2.8-mile course, racing for one hour. Eddie put the TR-6 through its toughest race against some of the roughest competition equipment in the country. He finished the race with a three-mile lead, and lapped every rider at least once!

At Ensenada, Mexico, one week later, after being tangled up in a starting line crash which de-railed the rear chain, Mulder started the event four minutes late. He finished in second place, in a performance even better than the Corriganville show. Plans were made to run the bike at the drags and then write this article for CYCLE WORLD, but getting married and a honeymoon in Europe delayed the project a full year. Last year Mrs. Owens and Pat rode the bike in the William Johnson Memorial Road Sports Rally for the second year in a row before it took on more competition miles.

"While in street trim, I rode it at the Fontana (California) drag strip and won the 'B' street class that day, turning 97.93 mph. In the run-off for top street eliminator it turned 100.52 mph!

"Next came another Corriganville, now called Hopetown since Bob Hope took up motorcycling and bought it, where our hero Mulder did it again, making it two years in a row on the same machine!

"Best part of this success story is that the engine was 100% standard when Eddie won the first race on it in 1963. And now, three years later, the engine has not been removed from the frame. Meaning, of course, all of the major engine components are still in place. Additional performance was gained by modification of the cylinder head, pistons, carburetors, lightening and polishing of pushrods, rocker arms and tappets, generally increasing the rpm range. Introduction of the 1966 'B' range machines saw factory use of many of the modifications proven so successful here." •

DRAG RACING might not be everyone's "cup-o'-tea" and after trying the real thing we can see why. The 10- to 12-second trip up the tarmac, on some occasions, can have all the excitement of a scramble, dirt track and road race rolled into one very short period of time. Although considerable skill is involved, there is no substitute for enough courage to take engine revs to the redline, "pop" the clutch and "light off" the rear tire. What happens from that point is anyone's guess.

In England, this form of racing is referred to as sprinting and has become very popular. While it is still looked upon as an American sport throughout the world, some major contributions to drag racing have come from England in recent years. It must be admitted that the Avon "slick" has gone a long way to lowering times through the quarter-mile. The slick is a treadless, flat section rear tire and is made from the high hysteresis rubber that has become so popular in road racing. Avon pioneered the

CYCLE WORLD
ROAD TEST

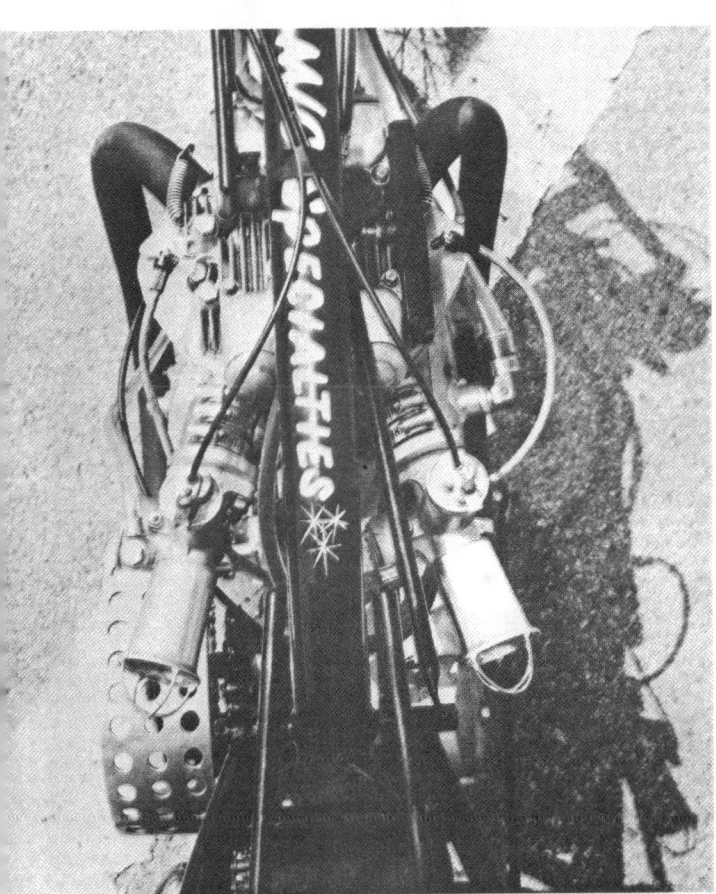

HAGON-TRIUMPH DRAGSTER

would continue, in limited quantity, of the precious slicks for dragsters.

One of the foremost exponents of sprinting in the land of tea and crumpets is a small, bespectacled, soft-spoken man of many talents. Alf Hagon is his name. For many years he was unbeatable in grass-track racing, which is a blend of motocross and dirt track. In the early days, Alf worked and raced for Tom Kirby, and perfected the Kirby-Hagon grass bike, which established an entirely new concept in design and is used by virtually everyone in grasstrack racing today.

When Alf retired from grass-track and took up drag racing, he soon started to apply his mechanical skill to the new sport. He found that a derivation of his successful grass-track machine could, with everything else being equal, give better quarter-mile times than existing equipment.

Hagon dragsters have become so popular in England that Alf is now producing and selling a kit, which is distributed in the U.S. by Kosman Motorcycle Specialties, Daly City, Calif.

The frame is constructed of seamless, mild steel tubing, and all joints are brazed. A large two-inch diameter top tube and a smaller rear down tube carries two quarts of engine oil. The "seat" is brazed to the rear sub frame, and this assembly is bolted to the main frame, very similar to the older rigid and sprung hub model Triumphs. Steering head and front down tubes are rather small compared to more conventional motorcycles. However, on this type of machine there is very little going on in the way of stress loadings in the front end. For one reason, seldom is the full weight of the front end acting on the wheel when the machine is running; and secondly, there is no front brake to impose its loads.

In fact, in drag racing, weight is probably more critical than in any other form of motorcycle competition, as the whole object of the game is to move weight from rest through a given distance.

This accounts, too, for the unusually small front forks

development of this unusual compound, which is sometimes called "dead" rubber, because it has so little life or elasticity, but allows a great deal of traction. In fact, this compound is credited with doing far more to decrease lap times than horsepower. Today, there is more racing than ever before but fewer accidents, as a result of this strange material. When Avon decided to quit building racing tires of any sort, it was a shock to the racing world, but reprieve came later when they announced that production on this type machine. Originally, as supplied by Hagon, the machine had Alf's own rubber band fork configuration, but Kosman fitted a pair of Yamaha 60 fork tubes in upper and lower fork crowns of his own manufacture.

We had hoped to try the Hagon theory and use a standard 650cc Triumph engine to compare quarter-mile times with a conventional motorcycle, but some tuning had been done on the engine. This consisted of Precision Machining drag cams, 11:1 Triumph pistons and Wert's aluminum push rods. Otherwise it was a standard 1962 pre-unit construction Triumph engine. Some indication of what can be expected may be gathered from the fact that the machine broke the quarter-mile 650cc gas record on its very first outing. This was at Fremont Drag Strip, where it turned 11.3 seconds and 117.95 mph and is recognized as a new national record.

During our testing we were not as fortunate, as magneto trouble had developed since the Fremont runs. To say this made things exciting would be an understatement. It was anyone's guess as to when the power was going to come on! On occasion, things would happen according to plan in first gear, then momentarily die in second, only to have the power suddenly come back, picking up the front end.

Despite all the guess work in throttle response and mis-firing, even on the best runs, the times we obtained proved the Hagon layout has considerable merit. A very low overall weight makes it an extremely feisty device, requiring much more talent than its roadster counterpart; but then it is designed for a specific purpose, and this is usually what sorts the men from the boys in motorcycle racing machinery. ■

HAGON-TRIUMPH DRAGSTER

SPECIFICATIONS

List price	P.O.R.
Suspension, front	telescopic forks
Suspension, rear	none
Tire, front	2.00-18
Tire, rear	Avon Slick-18
Engine type	twin, ohv
Bore and stroke (inches-millimeters)	2.79 x 3.23, 71 x 82
Displacement (inches^3-centimeters^3)	40-650
Compression ratio	11.0:1
Carburetion	(2) 1 3/16" Amal monobloc
Ignition	magneto
Bhp @ rpm	55
Oil system	dry sump
Oil capacity, pts.	4
Fuel capacity, gal.	1.5
Starting system	run and bump

POWER TRANSMISSION

Clutch	multi-disc, dry plate
Primary drive	single-row chain
Final drive	single-row chain
Gear ratios, overall:1	
5th	none
4th	5.25
3rd	6.15
2nd	9.14
1st	12.60

DIMENSIONS

Wheelbase	57.8
Saddle height	23.8
Saddle width	4.5
Footpeg height	13.2
Ground clearance	2.8
Curb weight	207

PERFORMANCE

Top speed (with test gearing)	117
Maximum speed in gears @ 8,000 rpm	
5th	none
4th	117
3rd	100
2nd	68
1st	49
Mph per 1000 rpm, top gear	14.6
Standing 1/4-mile, sec.	12.3
terminal speed	105

BY CHRIS LAVERY

THE SIGHT OF TWO TRIUMPH ENGINES coupled in a single frame is nothing new. You see it on the drag strips repeatedly. But just look at this Norton-Triumph; the crankcases are welded together!

Sorry you can't see this monster in action, but right now it's in the window of Bob Foster's emporium in Parkstone, Dorset, England. For those who don't know, Bob was world's 350cc road-race champion in 1950 on an overhead-cam Velocette.

This 61-cubic-inch Norton-Triumph vee-four is not Foster's, however. It belongs to Arthur Harris, an English scrambler who bought it in Napier, New Zealand, from Bill Plummer, the man who built it.

Some time back, Plummer was NZ sidecar racing champion, and he dreamed up this unusual method of joining two Triumph engines, because it was the only way he could squeeze them into a Manx Norton frame without too much alteration.

The scheme worked out remarkably well. Framewise, all Bill had to do was weld in some extra-wide down tubes to clear the front cylinder head, then find a new location for the oil tank.

Of course, the rear cylinder block and head had to be reversed to make room for the carburetor. Forward-facing carbs, as a means of free supercharging, are a myth. At somewhere around 80 mph, the air pressure over the jet builds up to the same level as that in the float chamber (14.7 lb.sq.in.), and so the gas stops flowing.

But the rear carb on this plot is probably well enough shielded from a direct air blast to continue to work normally. Special manifolds were made to offset the air intakes relative to one another.

Before welding, the crankcases were cut away at an angle just clear of the flywheel rims. The cases are joined in such a way that the included angle between banks is about 60 degrees.

However, the crankshafts are coupled so that each pair of pistons is at top center, while the others are at bottom center. Thus, the firing intervals are evenly spaced, and an ordinary four-cylinder distributor is used on the front crankcase.

Although the two crank chambers are in communication with one another, separate oiling systems are used, fed from a common tank. To join the shafts there is a train of gears in a case outboard of the rear engine sprocket. From there the drive goes to an AJS 7R gear box.

What use is a machine like this for European scrambling? None at all. Then why did Arthur Harris buy it? Because, after begging a ride on it during a visit to New Zealand, he just couldn't resist it.

Although it should be good for 160 mph, it is probably too short for dragging. But with its 75 or 80 bhp, it should be capable of creating a stir in sidecar racing.

■

TRITON-VEE FOUR

No Rest For the Winner
TRIUMPH 500 ROAD RACER

THAT A TRIUMPH REPOSED in the winner's circle at Daytona last year was not an earth-shaking situation; it has long been a widely — and wisely — accepted fact that the Meriden Works twins are capable of some pretty great things. The win, however, was a bit of a surprise, in view of the show of strength by the ever-quick, home-built H-Ds that were favored in the 200-mile climax to the annual Florida speed orgy. Mild-mannered Buddy Elmore, who is, in reality, Superracer, came from deep in the pack at the start and plucked off the front running dicers one by one until the day had become his, and his name was duly entered in the record book. Very little was known about Buddy and his Triumph road racer before Daytona. After Daytona, there was little point in taking a detailed look at Buddy or his machine; the essential story in racing is not the man nor the machine, but the race, itself.

Now, nearly a year later, with last year's Daytona just so many wonderful memories, and anticipation building for this year's running of the classic, we find ourselves delving into an analytical look-see at what Triumph will offer for the coming contest. The machine we are about to view, unlike Buddy's Triumph-Baltimore mount, is owned by Johnson Motors, west coast distributor for Triumph motorcycles in the U. S. The bike, differing little from the east coast version, has been most capably modified and maintained by JoMo's Dale Martin whose conversance with engineering theory and articulation in presenting the story behind the bike made our task of prying a most pleasant and rewarding experience.

The chassis of the Triumph road racing machine has all standard dimensions. It sports gusseting at the swing arm pivot, which has been incorporated on the '67 production frames, and additional gusseting on the arms where they tie into the crosspiece. The large-diameter top and down tubes are the same as last year's frame — another mod incorporated on the production frame for this year.

Hubs and brake components are current Triumph items, and the only non-standard pieces used are the American-made racing linings. The dust cover on the front brake has been windowed to aid cooling and the rear dust cover has been removed to facilitate sprocket changes. In addition, the front brake lever arm has been lengthened for better mechanical advantage. The hubs are laced into 19-inch alloy rims that are fitted with a triangular section Dunlop KR-76 road racing tire in front and a K-70 Gold Seal in back. Martin plans to experiment with a KR-76, mounted on a WM3 rim, for the rear; the triangular section tire on the current WM2 rim has too abrupt a transition from the straight-up to the heeled-over position.

The 500cc T100R engine incorporates numerous modifications, but most of them have been made in the interest of longevity rather than power. The cast-iron cyl-

inders have been overbored the AMA-allowed .040 inch and fitted with replacement pistons which are sand cast and slightly heavier than the original permanent-mold ones. The standard crank and rods were polished and the rotating and reciprocating masses were then rebalanced to Triumph's standard 70-percent factor. The timing side main bearing and the rod bearings are standard. A single-lip roller bearing has been fitted to the drive side of the crank to permit crankcase pressure pulses to be transferred into the primary case, where they are damped by the additional volume. Three 1/16-inch holes were drilled through the wall separating the primary drive chamber from the crank chamber, and maintain proper oil level in the primary chamber by returning excess oil to the crank chamber, where it is picked up by the scavenge. This modification does away with the timed breather (whose purpose is to damp the pressure pulses), and the engine now employs an atmospheric breather. The lubrication system has been further improved through the incorporation of a large-capacity alloy tank (developed by Triumph-Baltimore) and the addition of a Corvair oil cooler mounted on the right side of the front down tube.

The cams used in the Daytona road racer are factory racing items mounted in Torrington needle bearings. The cam lobes work against enlarged tappet bases that are accommodated by larger-than-standard bores in the barrels. Vapor oiling of the tappets replaces the pressure feed, and has been found to be completely satisfactory. The timing gears have been drilled and narrowed.

The combustion chamber shape, passed on to the roadster, is hemispherical; however, the road racer has inserted valve seats, while the roadster employs seats that are cast in. The design permits a two-degree reduction of included valve angle which affords a better valve-to-rocker relationship and increased lift and overlap for better cylinder filling. The intake valves are off-the-shelf JoMo items and the exhaust valves are pneumatic. JoMo valve springs, with aluminum collars which have a steeper angle in the keeper groove than the standard ones, and standard push rods are used. The standard rockers, which have been polished, feature adjusters that are shortened and lightened by having the square end ground off and then drilled with a screwdriver slot added, and the lock nut replaced with a connecting rod nut.

Fuel is supplied by two 1-3/16-inch Amal GPs, mounted to rubber sleeves with Bonneville adapters that have their threads turned off. The carburetors share a single-point-suspended float chamber and employ standard Amal intake bells.

Ignition is provided by a standard energy-transfer rotor-stator combination with the stator encapsulated in plastic. The alternator lead exits from the front of the engine, and a Lucas contact breaker assembly, which runs on its own bearings, is mounted on the right engine cover.

The transmission, like much of the machine, is made up of available Triumph parts. Basically, only two modi-

CYCLE WORLD
ROAD TEST

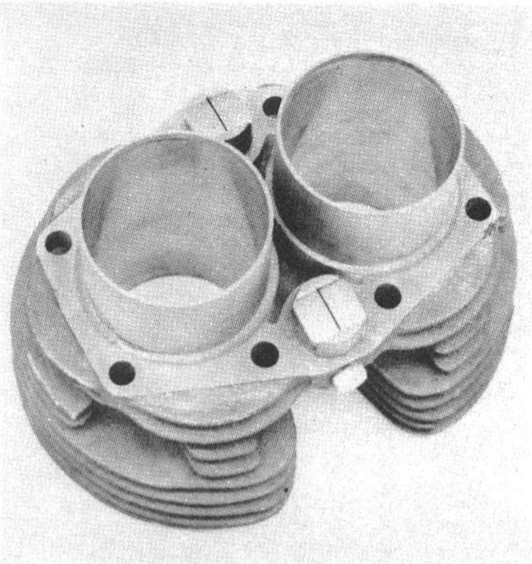

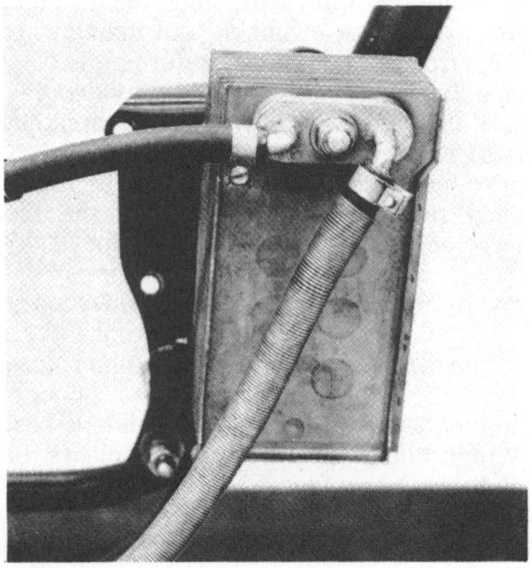

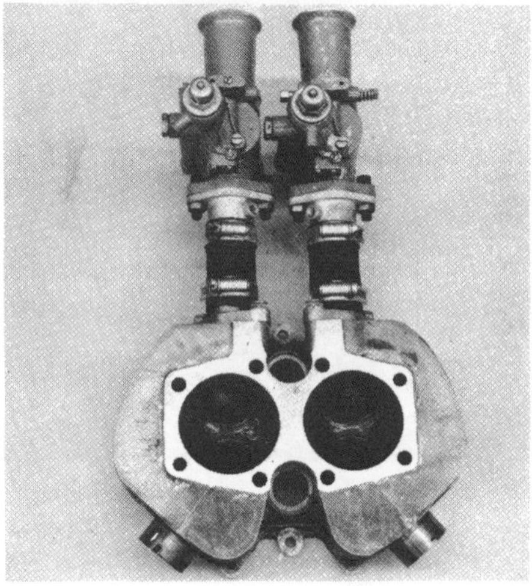

76

fications have been made to the assembly; the countershaft sleeve bushing on the mainshaft has been shortened and the sleeve is mounted on a roller bearing. The kick starter mechanism has been removed and the hole in the cover plugged. The rest is all standard, including the shifting quadrant which produces a shift pattern that is the reverse of Triumph's with the shifting lever turned 180 degrees. The stock clutch (58-tooth wheel) is fitted with steel plates and JoMo's relined friction plates and is connected to the 26-tooth engine sprocket with a standard primary chain with a riveted master link.

The short, fat fuel tank is something else. This item, manufactured by Triumph's sister company for production road racers, severely hampers movement of the rider's forearms and is about as handy as riding with an overinflated beach ball in one's lap. In addition, the tank ends about eight inches forward of the seat — a needless space that could very easily accommodate the volume of fuel that makes the current tank so fat. The matter is clearly a case of an interim solution and we presume Triumph will see fit to design another vessel for this chassis.

The handling characteristics of the Daytona road racer are quite good — perhaps not as good as a couple of other "legendary" production road racers — but the machine is responsive and handles confidently. Chassis flexing is nil and cornering line is constant and predictable. The brakes are good, despite the hard use they see. Braking could be improved, however, as is often the case with production units used for competition, to offer an additional margin for emergency situations or the increased braking activity required by very tight, twisty courses.

Without a doubt, the Triumph is one of the swiftest road racers ever built. Not only does the bike have a most respectable and competitive top end, but it accelerates hard and comes out of turns like gang-busters. It goes without saying that Triumph is working toward and hoping for a repeat of last year's Daytona victory. And with this machine as evidence of their abilities and determination, there is little question that Triumph will be keeping everybody honest at the 200-miler. ■

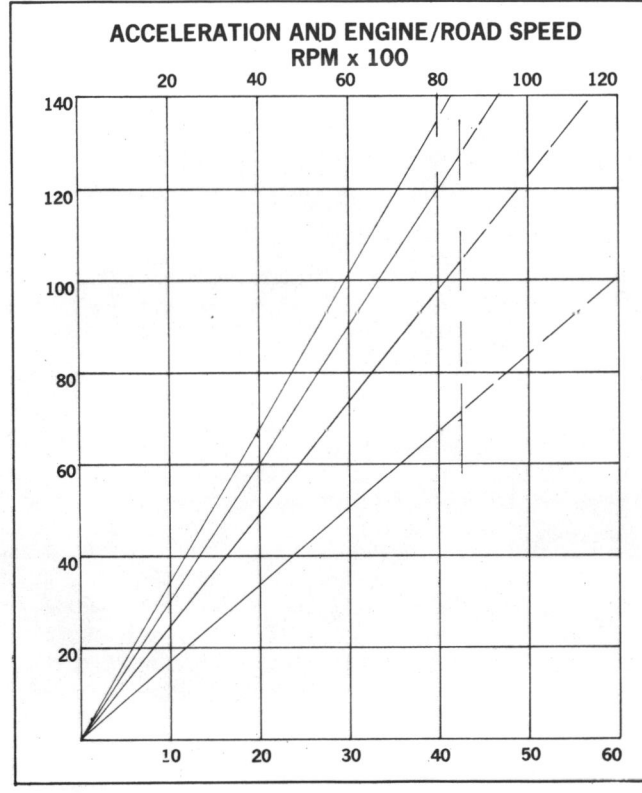

TRIUMPH ROAD RACER

SPECIFICATIONS

List price	n.a.
Suspension, front	telescopic fork
Suspension, rear	swing arm
Tire, front	3.00-19
Tire, rear	3.50-19
Brake, front	8 x 1.63
Brake, rear	7 x 1.13
Total brake swept area, sq-in.	65.8
Brake loading (test weight/swept area) lb/sq-in.	7.2
Engine type	ohv, vertical twin
Bore and stroke (inches-millimeters)	2.72 x 2.58, 69 x 65.5
Displacement (inches^3-centimeters^3)	30, 490
Compression ratio	9.5
Carburetion	Amal GP (2), 1-3/16"
Ignition	AC magneto
Bhp @ rpm	n.a.
Oil system	dry sump
Oil capacity, pts.	16
Fuel capacity, gal.	5
Starting system	push
Lighting system	none
Air filtration	none
Clutch	multi-disc, wet plate
Primary drive	duplex chain
Final drive	single-row chain
Gear ratios, overall:1	
5th	none
4th	4.2
3rd	4.7
2nd	5.7
1st	8.4
Wheelbase	53.5
Seat height	30.5
Seat width	11.0
Foot-peg height	14.0
Ground clearance	7.0
Curb weight (w/half-tank fuel)	319
Test weight (fuel and rider)	479

PERFORMANCE

Top speed (with test gearing)	114
Maximum speed in gears (@ 8500 rpm)	
5th	none
4th (8000 rpm)	134
3rd	127
2nd	104
1st	71
Mph per 1000 rpm, top gear	16.8

78

PHOTO DAVID GOOLEY

Something for the Mojave Bailiwick

A PAIR OF DESERT SLEDS

The word, "sled," as applied to a motorcycle, means desert racer and probably has origin in the delightful scraping sensation one feels as the skid plate hits ground. But the connotation goes even farther in that the term brings to mind, more often than not, a throbbing big twin from Coventry decked out in appropriate accoutrement.

As the riding gets faster each year and the demands on machinery become more grueling, a desert-ready bike is now a far cry from most machines on the showroom floor. Thus, one old pro, Ted Lapidakis of the Checkers M/C, decided to bring the two closer together. The result is that the reader may, without lifting a wrench, buy a brand-new 500cc or 650cc Triumph desert sled ready to race (after a few break-in miles, naturally).

The price tag is a bit more ($1564 for the 650, $1430 for the 500), but riders who bought a new Triumph and tried to duplicate the work done would receive a rude fiscal shock, unless they were lucky enough to get discounts on the parts and free labor.

Modifications to both bikes are the same with one exception: while the 650's fork angle remains standard, the 500's frame has been jacked out slightly for greater rake.

Starting material for both sleds are the single-carburetor TR6C and T100C Triumphs. Tires are changed for 4:00 x 19 Trials Universal up front and 4:00 x 18 Dunlop Sports rearward, the standard traction for desert. Front fork travel is extended 1-1/2 inches and heavy-duty dampers of Lapidakis' own construction replace the originals. A wider, tougher stainless steel front fender goes on. The center strut of three steel struts holding the fender on is heavier than the others; tied down at two spots several inches apart on each fork leg, it serves as a very rigid U-brace.

Rear suspension modifications are less extensive, involving only the installation of extra-long 13-inch Girling dampers. The tail of the rear subframe, however, is heated and bent upward to provide more room for wheel travel. An alloy rear fender replaces the stock item.

Then, footpegs undergo modification, receiving bracing. A shield between each peg and the chaincase and timing covers is installed with an eye to preventing possible holing of the covers by the pegs in the event of a fall or shunt with a boulder. Of course, the frontal crankcase area is protected by the time-honored oversized skid plate.

Breakage protection is also extended to the spokes, which are wired at their intersection points to prevent loose or broken ones from flapping about or puncturing inner tubes. On the rear wheel, a rather subtle precaution — that of tackwelding the hub to the wheel flange — prevents spoke movement under heavy pounding, and subsequent breakage.

Very little is done to the engines, either the 500 or 650, the theory being that there is already enough power in the standard units and that reliability rather than the last smidgin of bhp is the determinant of long-distance winners. Indeed, reliability is the object of most of the mods, even that of the specially fashioned straight-through exhausts, which are tucked out of harm's way better than the stock variety.

Lapidakis rewires the electrical system, enclosing wires in thick plastic tubing, and installs soft rubber sparkplug covers, which seem to hold better. Square nut Webco valve covers replace the stock ones and are drilled and wired so they'll *stay* on. In addition to a float bowl extension for smoother fuel flow, the carburetor receives a rubber cover to keep dirt out and the cap on. A heavy duty "Q" air filter is installed. Finally, a shorter kick starter lever, thought to be easier to use and provide more cranking speed when the rider stalls in the boonies, is mounted.

Another modification rids one of the problems in getting to the oil tank filler when a Bates desert seat is mounted: The spout is relocated. Lapidakis then punches a hole in the rear downtube, through which the oil spill tube runs, thus eliminating another external line to get fouled.

Other additions consist of overlay wheel sprocket and beefier handgrips with dust protectors.

In short, these modifications add up to a competition-ready machine on the showroom floor, a replica of one that Lapidakis rides and gives a good showing on every time he goes out. ■

BONNEVILLE TRIUMPH

CONTINUED FROM PAGE 37

and that again we would have to be quite cautious about applying full throttle at much less than 7000 rpm. But, as experience proved, the pistons were even better than we had thought: at no time was there even a trace of detonation, and on our first tune-up run we were running too lean and with too much ignition advance — a sure invitation to violent detonation.

Down at the engine's lower end, roller bearings were used for the main journals, and the crank/flywheel assembly rebalanced to a 75% factor. Stock rods were employed with stock insert-type bearings and stock clearances at the rod journals. Webco's cylinder base gaskets were used to shim the block up for the necessary .060" squish clearance. These gaskets are, if you haven't already heard, of a flexible, yet non-compressing material and do the job nicely when you are trying to "stack" to close tolerances on the upper end of an engine.

Photos of the engine's "timing side" will show one major modification made: the installation of a Parilla crankcase breather. This was threaded into a boss welded on the timing case, and it allows the crankcase to breathe completely free, rather than chuffing through the standard timed breather. There is something to be said for timed breathers, but we are of the opinion that there is a slight gain in power when the crankcase is completely vented to the atmosphere. Also, experiences others have had indicate that the "free" breather tends to make the engine slightly more oil-tight.

In addition to these things, there were many small refinements: for instance, the top of the crankcase, where the cylinder block bolts down, was machined "square" with the crankshaft axis. Also, the cylinder block itself was machined true on both faces and overbored .020 to make the bores parallel and perpendicular to the crank axis. The timing gears' tooth faces were lapped by running them with a coating of very fine grinding compound, which was, of course, carefully cleaned away before final assembly of the engine. And, naturally, the engine also has had other little fine touches, such as lightened and polished rocker arms.

The various features of this modified engine were decided after a series of conferences between ourselves and Gary Robison of Harman & Collins. Robison knows a great deal about the Triumph engine (as evidenced by the success he has had with his TT bikes) and no matter how much you fancy that you know about engines generally, there is no substitute for specific knowledge when trying for maximum power from some particular engine. For the same reasons, and because of the acute shortage of time, Gary was also given the thankless task of translating our conversation into a running engine — which he did, and very nicely, too. We also had help from another "outsider", Brian Crawford, who maintains and tunes the technical editor's road racing bike. Brian worked many long nighttime hours on the Triumph, and without his assistance we might never have reached Bonneville at all.

The final "different" touch on our Bonneville Triumph was the installation of a complete Honda ignition system. We anticipated having to experiment considerably with spark advance, and it would have been far too much trouble to re-time Triumph's energy transfer system with every change of ignition advance. The Honda battery and coil setup (we ran without a generator, of course) will deliver at least 25,000 sparks per minute without a miss, and with a fully charged battery it can continue this for about 8 hours. That was, obviously, more than adequate for our purposes.

As we have said, nothing much was accomplished by us at Bonneville this year. It is not possible to sort out the tuning on a new engine in 4 runs, and about all we did was to (a) prove that Webco pistons did in fact give detonation-free running with 12:1 compression ratio and (b) discover how bad the salt can get when the weather is wrong. The salt was extremely wet, with the water-table barely an inch below the surface, and this also left it very soft and the cars quickly cut ruts all the way down the course. There was one spot so soft and rutted that you could actually feel the bike sink slightly and when this occurred the engine would begin to drop revs. At all times there was a fantastic amount of slippage, and our Hero Rider reported that he could send the revs soaring just by sliding forward on the seat and taking a bit of weight off the rear wheel. We were, as a matter of fact, turning enough revs to be running 145 mph when the timing lights told us our speed was only 136 mph.

Team-mate Rich Richards, who has been running his 40-inch "fueler" a long time and knows how to get it tuned with a minimum of experimentation, was in relatively good shape. His son, Gary, was able to get it through with a two-way average of 159.222 mph to establish a record for class PS-A 650cc bikes, but this was virtually the same speed that the same bike ran three years ago when Richards and Son captured the A-A 650cc record — with no fairing at all. Gary Richards tells us he had well over 8000 rpm showing on the tach during his runs, which should have given a speed of around 190 mph — which means that the rear wheel was slipping to the tune of 30 mph.

So, the CYCLE WORLD Team was successful — and at the same time it was not. We retained our old record, but were unable to improve upon it; and Rich Richards collected a new record with his bike. If the salt is good in 1965, perhaps we shall do something more spectacular. In the meantime, we will be running our PS-C 650cc Triumph on a dyno, to get it in proper tune and to see just how much horsepower it will deliver. We will give a report of the dyno work when it is completed. •